Olive Oil - New Perspectives and Applications

Edited by Muhammad Akram

Published in London, United Kingdom

IntechOpen

Supporting open minds since 2005

Olive Oil - New Perspectives and Applications
http://dx.doi.org/10.5772/intechopen.87514
Edited by Muhammad Akram

Assistant to the Editor: Rabia Zahid

Contributors
Didar Üçüncüoğlu, Dritan Topi, Gamze Guclu, Hasim Kelebek, Serkan Selli, Ana Florencia de Alzaa, Leandro Ravetti, Claudia Guillaume, Aadil Bajoub, El Amine Ajal, Rachid Nejjari, Salah Chaji, Sanae Moussafir, Abdelmajid Soulaymani, Isicio Ortega- Medina, Juan José Soto-Castillo, Anita Trajkovska-Broach, Anka Trajkovska Petkoska, Mohammed El Haouari, S. Vimala Devi, K.B. Sridhar, Thangamani Dhandapani, Muhammad Akram

First published in London, United Kingdom, 2022 by IntechOpen
IntechOpen is the global imprint of INTECHOPEN LIMITED, registered in England and Wales, registration number: 11086078, 5 Princes Gate Court, London, SW7 2QJ, United Kingdom
Printed in Croatia

British Library Cataloguing-in-Publication Data
A catalogue record for this book is available from the British Library

Additional hard and PDF copies can be obtained from orders@intechopen.com

Olive Oil - New Perspectives and Applications
Edited by Muhammad Akram
p. cm.
Print ISBN 978-1-83968-414-2
Online ISBN 978-1-83968-415-9
eBook (PDF) ISBN 978-1-83969-251-2

For EU product safety concerns:
IN TECH d.o.o., Prolaz Marije Krucifikse Kozulić 3, 51000 Rijeka, Croatia, info@intechopen.com or visit our website at intechopen.com.

Meet the editor

Dr. Muhammad Akram is an assistant professor in the Department of Eastern Medicine, Government College University Faisalabad, Pakistan. He received his Ph.D. from Hamdard University Karachi-Pakistan in 2013. Dr. Akram was a chairman in the Department of Eastern Medicine and Surgery, University of Poonch, Rawalakot Azad Kashmir from 2015 to 2017. He received many honors and awards during his career. He serves as an editor and invited reviewer of several national and international journals. He has numerous publications and presentations to his credit, and he is an active member of several professional societies. Dr. Akram's research interests include hyperuricemia, xanthine oxidase inhibition by some selected medicinal plants, enzyme inhibition, Indusyunic medicine, phytochemistry, poisonous plants, bioactivity, and phytopharmaceutical evaluation of herbal drugs and their natural products, biochemistry, and bioinformatics.

Contents

Preface

This edited volume is a collection of research chapters on olive oil and its applications. The book includes scholarly contributions by various authors and is edited by a group of experts pertinent to the agricultural, medical, and biological sciences field. Each contribution comes as a separate chapter complete in itself but directly related to the book's topics and objectives. Over nine interdisciplinary chapters, the book discusses the latest research in the production of olive oil and its applications in health and nutrition. This book provides advanced knowledge for researchers in related fields, students, life science researchers, and interested readers.

Dr. Muhammad Akram
(Associate Professor) Chairperson,
Department of Eastern Medicine,
Government College University Faisalabad,
Pakistan

Chapter 1

Health Benefits of Extra Virgin Olive Oil

Anka Trajkovska Petkoska and Anita Trajkovska-Broach

Abstract

Extra virgin olive oil (EVOO), also called the "*Elixir of the youth and health*" by the Ancient Greeks, is a cornerstone in the Mediterranean diet, which has been recognized as one of the healthiest and most sustainable dietary pattern and lifestyle. In this chapter, a brief overview of the major and minor components of EVOO is given followed by a review of their health benefits. In particular, the antioxidant, anti-inflammatory, anti-cancer and cardiovascular protective effects of EVOO are emphasized. At the end of this chapter, the reader would benefit by realizing that EVOO, as a functional food, proves the Hippocrates's quote "*Let food be thy medicine and medicine be thy food*".

Keywords: Extra virgin olive oil, Mediterranean Diet, health benefits, antioxidative effects, anti-inflammatory effects

1. Introduction

The cultivation of olive trees and production of olive oil have been related to the history and the culture of the people, who lived around Mediterranean Sea. A lot of archeological evidence has witnessed that inhabitants in this region consumed olives since the copper age. Over the millennia, olive oil has been used not only as a dietary ingredient, but also as lamp fuel, for cosmetics and pharmacological uses, for special rituals such as anointing royalty, warriors, etc. Greek philosophers have showed interest to examine its nutritional and medicinal benefits, while Aristotle and Hippocrates have recommended the olive oil for treatment of many diseases, such as dermatitis, stomach and intestine problems, as well as sun protection and burnt skin [1–4].

Nations with different ethnic, historical, and cultural backgrounds and religious beliefs have lived in the Mediterranean region, where olives have been cultivated [5]. Over the past several decades, experts are recognizing that the diet of the people in the Mediterranean is among the healthiest diets in the world, because of the lower incidences of many chronic diseases, like cardio-vascular and age-related diseases among these people [6–8]. In particular, in 1960's, the phrase *Mediterranean Diet (MD)* was coined by Ancel Keys based on his research of the traditional culinary practices in the rural areas of Southern Italy, Crete, and other countries in the Mediterranean basin. His research, which started as a part of the Seven Countries Study with a special focus directed to the nutritional practices for the prevention of coronary heart diseases by replacing saturated fats (saturated fatty acids, SFA) with unsaturated fats, including polyunsaturated fatty acids (PUFA) and monounsaturated fatty acids (MUFA), mainly sourced from the olive oil, showed that people living in the Mediterranean basin indeed have less cardiovascular events

Figure 1.
Mediterranean diet pyramid (© 2009 Oldways preservation and exchange trust).

and reduced risk of other diseases [1, 7, 9–11]. Half century later, in 2010, MD even became a part of the UNESCO's intangible cultural heritage, where it is defined as "*a set of skills, knowledge, rituals, symbols and traditions concerning crops, harvesting, fishing, animal husbandry, conservation, processing, cooking, and particularly the sharing and consumption of food*" [1, 5, 12].

The traditional Mediterranean cuisine emphasizes an abundant intake of variety of vegetables, fruits, legumes, whole grains, and nuts, moderate intake of seafood and fish, and a low consumption of red meat, sweets and other processed food, while red wine is usually consumed with the meals. Olive oil (OO), known as the *liquid gold,* and mainly used as a virgin olive oil (VOO) or as extra virgin olive oil (EVOO), is the hallmark of the MD [12, 13]. It is, in fact, the main source of healthy fats in the diet, which along with the other OO bioactive components, such as polyphenols, are often associated to the longevity, well-being and a lower incidence of chronic diseases, particularly cardiovascular diseases (CVDs) in the populations living in the Mediterranean region [9, 14, 15]. **Figure 1** below is a schematic presentation of MD with the foods consumed in abundance at the bottom of the pyramid and those consumed rarely at the top of the pyramid. Important part of MD is the social aspect of the Mediterranean lifestyle, conviviality and daily physical activity, also presented at the bottom of MD pyramid.

2. Composition of olive oil

The *major* components of OO are saponifiable lipids (~98%), which are mostly *triglycerols*, MUFA (oleic acid, 55%–83%), PUFA (linoleic acid, 2.5%–21%) and SFA (palmitic acid, 7.5%–20%). *The minor* (unsaponifiable) fraction (~2%) has more than 200 chemical compounds among which more than 30 are *phenolic compounds*, such as hydroxytyrosol, oleuropein, oleocanthal and tyrosol, phenolic acids (vanillic acid, syringic acid, gallic acid, etc), flavonoids (eriodictyol, apigenin, luteolin, etc.), secoiridoids (oleacein, oleocanthal, etc.) and lignans [8, 11, 16–23].

The lipid fraction (MUFA, PUFA) in OO gives its lipophilic character responsible for protective properties on coronary, autoimmune and inflammatory disorders, granting anti-thrombotic and regulation effects of blood pressure [24]. The minor components like α-tocopherol, tocotrienols, and carotenoids, such as β-carotene and lutein, squalene and other triterpenes, sterols, and pigments are also important for human health and they are responsible for the oil taste and aroma of olive oil. The chlorophyll (a non-oily component) in OO, determines the color of the OO that can be lost in the refining process. Its role is a facilitator of cell growth, and its part in the stimulation of the formation of blood cells (hematopoiesis) has been reported [3]. The bioactivity of the phenolic compounds is related to different properties, antioxidant and anti-inflammatory, although the molecular mechanism of these compounds in relation to many diseases could have different cellular targets [16]. Their content in OO varies depending on climate, cultivar, ripeness of the olives at harvesting, as well as the production process of the olive oil [14, 21, 25–30].

Different olive oil types are classified according to their acidity, expressed as amount of oleic acid (*International Olive Council (IOC) standard).* For instance, extra virgin olive oil (EVOO) has a free acidity of <0.8 grams / 100 grams, virgin olive oil (VOO) has a free acidity of <2 grams / 100 grams, and ordinary OO has a free acidity of <3.3 grams/100 grams. During the refining process some of the important components, like phenolic compounds and squalene, are lost; thus, *EVOO is the olive oil with the highest phenolic compound content* with the mean total polyphenol content of >55 mg/100 g [14].

Among the many constituents of olive oil, particular attention has been focused to the *phenolic compounds* due to their antioxidant effects, but also to their anti-inflammatory activities since that inflammation is an important etiologic factor for several non-communicable diseases [16, 24, 25, 31]. There are various OO phenolic compounds, which can be classified as simple phenols (hydroxytyrosol, tyrosol), secoiridoids (oleuropein) and lignans. Among them, *oleuropein, hydroxytyrosol and oleocanthal* are the most studied compounds with proven benefits, such as being strong antioxidant compounds counteracting reactive oxygen species' formation [18, 20, 32–35].

3. Oxidative stress and inflammation

Oxidative stress is defined as an imbalance between the oxidant and antioxidant systems of the body, in favor of the oxidants. It occurs when excessive generation of free radicals, reactive oxygen species (ROS) and reactive nitrogen species (RNS), produced during the normal cell metabolic processes or by external factors (pollution, smoke, radiation and chemicals) is unbalanced by the body's antioxidant defense system. The oxidative stress has been implicated in the etiology of many diseases; in particular, the oxidative stress is one of major cellular features in the onset of pathological conditions, such as neurodegenerative disorders (Alzheimer's and Parkinson diseases), renal disease, diabetes, ischemia, atherosclerosis, pulmonary dysfunction, cancers, and aging [3, 8, 25, 36–40].

The oxidative stress can activate a variety of transcription factors, which lead to the differential expression of some genes involved in different inflammatory pathways [38–41]. Inflammation and oxidative stress are interrelated and closely linked to many pathophysiological processes; one of them may occur before or after the other. In most of the cases, both of them take part in the pathogenesis of the chronic diseases [37]. The oxidative stress and inflammation, when unbalanced by the antioxidant defense system, can lead to changes of the crucial biomolecules in the body. Namely, ROS and RNS as highly reactive molecules can damage cell structures, such as the carbohydrates, nucleic acids, lipids, proteins and alter their functions through lipids' peroxidation, DNA damage and proteins' oxidation that lead to cell mutation, abnormal cellular growth, apoptosis, and necrosis (**Figure 2**) [30, 38].

On the other hand, the human *antioxidant defense system* includes endogenous and exogenous antioxidants. Enzymes, such as superoxide dismutase, catalase, glutathione reductase, glutathione and glutathione peroxidase, metal binding proteins form the body exogenous antioxidant system, then some elements or compounds (selenium, zinc, vitamins A, C and E, phenols), mainly derived from the diet, belong to the group of exogenous antioxidants. Therefore, the level and the diversity of antioxidants in the body are very important for *counteracting and neutralizing the oxidative stress*, which is critical for cell viability, activation, proliferation, and organ function. Enzymatic and nonenzymatic antioxidants are usually effective in blocking the harmful effects of ROS. However, in the human cells, *de novo* antioxidant production is very limited. Also, in pathological conditions, the antioxidant systems can be overwhelmed. Hence, the intake of external antioxidants can assist in fighting the oxidative stress. In this context, the role of *dietary antioxidants*, such as polyphenols, carotenoids, tocopherols, tocotrienols and others is very important. EVOO, as a rich source of antioxidants, has been recognized and recommended by many experts, medical doctors, dieticians, nutritionists to help in protection against the oxidative stress, inflammation and related diseases [25]. In this context, the European Food Safety Authority (EFSA) in 2011 has approved a health claim stating that the dietary intake of VOO's polyphenols is able to protect blood lipids from oxidative damage. However, in order this claim to be valid, 5 mg/day of hydroxytyrosol and its derivatives should be consumed (Commission Regulation, EU, 432/2012) [16, 42–44].

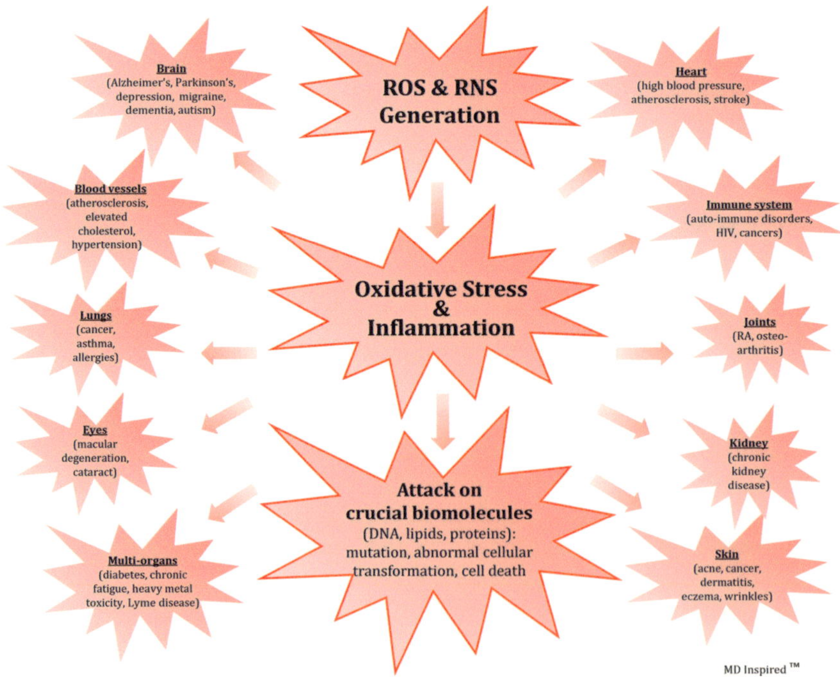

Figure 2.
Oxidative stress and inflammation and their effects on different organs.

4. Health benefits of EVOO

4.1 Antioxidative and anti-inflammatory effects of EVOO bioactive constituents

The benefits of OO, especially those of EVOO, in the human organism are well established, and they are mainly due to its composition. The predominant MUFA present in EVOO is the oleic acid. Also, the microconstituents, such as phytosterols, squalene, tocopherols, phenolic compounds, terpenic acid derivatives, etc. have particular beneficial effects on the human health and well-being. Among them, the phenolic compounds are the most studied and proven for their remarkable antioxidant activities. It is worth noting that secoiridoids and alcoholic phenols (hydroxytyrosol) are present in much higher amounts in EVOO compared to refined OO. These phenolic compounds are characterized by a broad spectrum of biological activities, such as reducing the morbidity, prevention and slowing the progression of diseases associated with oxidative stress, due to their antioxidant activity [2, 6, 8, 37, 39, 45–49].

The benefits of EVOO phenolic compounds, are closely related to their chemical structure, specifically due to the presence of hydroxyl groups. The action mechanism could be attributed to the electron donating ability of the hydroxyl groups and subsequent formation of intramolecular hydrogen bonds with the free radicals. In addition to the antioxidant effect, *i.e.* the direct scavenging of reactive species, the EVOO polyphenols' modulation of gene expression plays a key role in their anti-inflammatory properties. A strong scientific evidence supports the association of phenolic compounds with the prevention or reduced risk of diseases caused and characterized by oxidative stress or inflammation, such as cancers, digestive disorders, metabolic syndrome, obesity, atherosclerosis and CVDs [19, 24, 27, 31, 49–51]. Beside antioxidant and anti-inflammation effects, polyphenols are also responsible for antihepatotoxic, antidiarrheal, anti-ulcerous in the digestive system, anti-allergic, anthelmintic, anti-osteoporosis effects, but also have anti-bacterial and antiviral effects [3, 4].

It is worth mentioning that the EVOO phenolic compounds' antioxidant and anti-inflammatory effects are synergic in nature giving rise to the profound bioactivity of EVOO against chronic diseases and different pathologies, where the oxidative stress has been implicated as underlying mechanism (neurodegenerative, digestive disorders, cancer, and metabolic syndrome). In particular, a large body of scientific evidence supports EVOO's phenolic compounds having the following beneficial effects (**Figure 3**):

- Anti-cancer and chemo-preventive effects
- Cardiovascular protective effects
- Antibacterial, antiviral and antifungal activity
- Respiratory effects
- Endocrine effects
- Gut- and immuno-modulatory effects
- Neuro-protective activity, etc.

In the next section, the health benefits of EVOO through the biological activities of some of its constituents, in particular the major phenolic compounds, are reviewed.

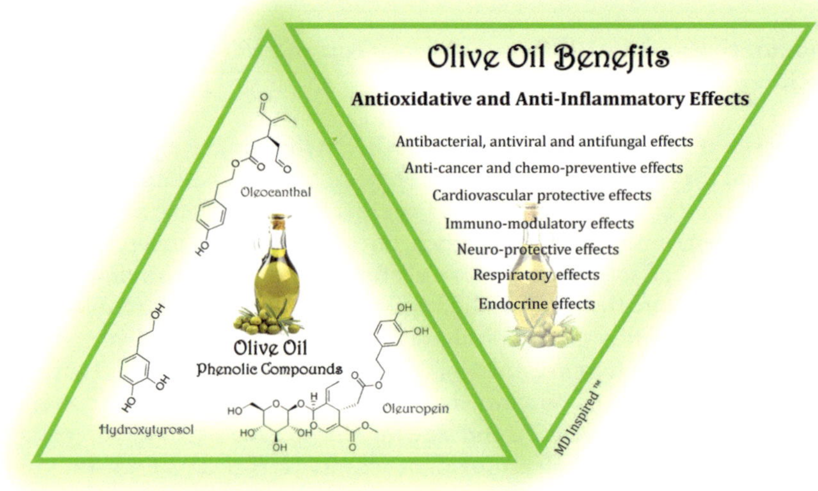

Figure 3.
Chemical structures of olive oil main phenolic compounds and the antioxidant and anti-inflammatory benefits of olive oil.

4.1.1 Hydroxytyrosol

As presented in **Figure 2**, hydroxytyrosol, oleocanthal and oleuropein are the most studied and most important polyphenols found in EVOO.

Hydroxytyrosol [14, 52–55] is considered to have the highest antioxidant and anti-inflammatory potency compared to the other EVOO's polyphenols. Its beneficial properties for human health are strongly related to the ability of the molecule to scavenge free radicals, ROS and RNS, as well as to activate endogenous antioxidant systems in the body. For instance, it is known, ROS are involved in the endothelial dysfunction contributing to atherosclerosis development. *In vitro* studies have shown that EVOO polyphenols are able to lower the oxidative stress and inflammatory-related sequelae associated with chronic degenerative diseases. It was found that hydroxytyrosol regulates the intracellular ROS levels in vascular endothelial cells and provides a molecular basis for the prevention of CVDs [56–58]. Another scientific evidence is relating hydroxytyrosol to a potential risk reduction for developing type 2 diabetes mellitus [22]. Its antioxidant activity has been shown to be efficient against oxidative damage *in vitro* in retinal pigment epithelial cells, which occurs in age-related macular degeneration lesions. Moreover, an *in vivo* assay revealed that oral supplementation of EVOO and specifically hydroxytyrosol reduces brain lipid peroxidation, acting as a powerful brain antioxidant [59]. The anti-inflammatory capacity of hydroxytyrosol has been also shown against acute ulcerative colitis. Namely, patients with inflammatory bowel disease are at increased risk for developing colorectal cancer. *In vivo* studies have showed that the diet enriched in polyphenols resulted in less incidence and multiplicity of tumors [60, 61].

Hydroxytyrosol in EVOO has also shown anti-HIV activity *in vitro*. It inhibited the viral integrase enzyme and the fusion of the viral envelope with host cells [58]. It has been considered as a potential microbicide, as well. In particular, more recent studies have shown that hydroxytyrosol has a potent antimicrobial activity against *Clostridium perfringens*, *Escherichia coli*, *Staphylococcus aureus*, *Salmonella enterica*, *Yersinia* sp., and *Shigella sonnei* [53, 62].

4.1.2 Oleocanthal

Oleocanthal [63–66] is ~10% of the total phenolic content in EVOO; it has pronounced antioxidant and anti-inflammatory activity and is responsible for the pungency of the fresh olive oil. Oleocanthal' anti-inflammatory activity is presented by inhibition of COX-1 and COX-2, cyclooxygenase enzymes, which catalyze important steps in inflammation pathways. It has been found that it has effects similar to Ibuprofen [8, 32, 65, 67, 68]. Although oleocanthal and ibuprofen are not chemically similar, they have similar anti-inflammatory properties that help reduce the risk of cancer and CVDs; oleocanthal was found to be even more potent than ibuprofen and is more effective in preventing inflammation. Other studies have been focused on its pharmacological effects, such as in cancer and neurodegenerative diseases, including the multiple myeloma cells, as well as activation of cytoprotective pathways promoting healthy aging [22, 40, 59, 69–75].

4.1.3 Oleuropein

Oleuropein is responsible for the characteristic bitter taste of unprocessed olives [1, 33, 76–78]; it has been found to perform a wide spectra of properties: antioxidant, anti-aging, anti-inflammatory, anti-atherogenic, anti-cancer, antimicrobial, antiviral, skin- and cardio-protective. Its hypolipidemic, hypoglycemic and neuroprotective activities have been studied, too [22, 76, 77].

Oleuropein was found to be effective against various strains of bacteria, viruses, fungi, as well as molds and parasites. Moreover, it also inhibits platelet aggregation. Oral treatment with oleuropein resulted in a reduced number of blood vessels proving strong anti-angiogenic properties to inhibit macrophage-mediated low-density lipoprotein (LDL). There are numerous studies confirming the anticancer activity of oleuropein, as observed in human cancer cell lines, such as breast adenocarcinoma, melanoma, urinary bladder carcinoma, colorectal adenocarcinoma, prostate cancer, lung carcinoma, glioblastoma, renal cell adenocarcinoma, etc [22]. Several experimental studies provide strong evidence that oleuropein exhibits beneficial hepatoprotective effects on the liver as the liver has been identified as one of the target organs of oxidative stress [76, 79, 80].

In the section below, the protective effects of EVOO in relation to particular diseases are briefly reviewed.

4.2 Protective effects of EVOO in different diseases

4.2.1 Cardiovascular diseases

Cardiovascular diseases (CVDs) often called "silent killers," are mostly common in urban community populations [13, 17]. Initially, the mechanism of EVOO as cardioprotector was based on the incidence on the so-called traditional risk factors (lipids and blood pressure), but the modern cardiovascular risk factors are extended to inflammation, oxidative stress, coagulation, platelet aggregation, fibrinolysis, endothelial function or lipids, or even modulation of the conditions, which are predispositions for CVDs, such as obesity, metabolic syndrome or type 2 diabetes melitus [81]. However, among the major CVD risk factors are still considered the hypercholesterolemia, hypertension, and atherosclerosis. CVD is a group of disorders such as chronic heart disease, stroke, rheumatic heart disease, peripheral arterial disease, congenital heart disease, pulmonary embolism, deep vein thrombosis and most of them proliferate in general by the accumulation of fatty deposits on the inner walls of the blood vessels causing a blockage of blood circulation to the arms, legs, brain, or heart [7, 17]. There is an evidence that EVOO's fatty acids play an essential role in the management of CVDs and do not cause deposits and blockage in the blood vessels [3, 4, 17, 82]. Current recommendations for primary prevention of CVD highlight the importance of dietary patterns including dietary sources of healthy fats, such as those high in unsaturated fat and low in saturated fat. EVOO is a perfect example as a dietary CVD preventor due to the presence of MUFA and PUFA [3, 13, 19, 35].

It is not only the unsaturated fatty acids in EVOO, but its phenolic compounds that have shown favorable results in modulation of oxidative balance markers of CVDs. These beneficial effects of EVOO were more pronounced in healthy patients compared to unhealthy subjects. EVOO benefits were also observed in insulin sensitivity, glycaemia, modulation of transcription of genes involved in lipid and glucose metabolism, inflammation, significant reduction of oxidized LDL, increased high density lipoprotein (HDL) cholesterol levels, etc. In fact, it has been shown that EVOO lowers the total blood cholesterol, LDL-cholesterol and triglycerides, while increasing the HDL-cholesterol level, which helps fend off the formation of fatty patches, thus stimulating the elimination of LDLs. Controlled clinical trials have shown that 1% reduction in total and LDL-cholesterol concentrations results in an ≈1.5% reduction in the incidence of CVD [3, 4, 83–87]. In addition, EVOO is capable of blunting oxidative stress by regulating the platelet oxidative stress and endothelial dysfunction. In humans, the role of EVOO as an anti-atherosclerotic nutrient is also supported by its ability to modulate expression of atherosclerosis-related genes in which LDL oxidation is involved. Another relevant mechanism of EVOO in modulation of CVDs is its anti-inflammatory effect in the vascular walls [88]. There are many long- and short-term studies that have shown that EVOO intake is also associated with a significant decrease in inflammatory markers, namely Thromboxane-B2 (TXB2) and Leukotriene-B4, which indicate EVOO's anti-thrombotic and anti-inflammatory activity in a postprandial state. Studies on subjects at high cardiovascular risk showed reduction in both blood pressure values, the systolic and the diastolic pressure [89, 90].

An inverse relationship between OO intake and coronary heart diseases' mortality and incidence has been reported within the EPIC (European Prospective Investigation into Cancer and Nutrition) cohorts [8], as well as during the PREDIMED (PREvención con DIeta MEDiterránea) study which included 7,216 men and women at high cardiovascular risk, aged 55 to 80 years [26]. The study linked EVOO and reduced risks of CVDs and mortality in individuals at high cardiovascular risk. More specifically, EVOO intake in the context of MD was associated with a reduction in the risk of CVD by 30% compared to controls subjected to low-fat-diet and reduced mortality in older high cardiovascular-risk individuals [91]. Estruch et al. have showed that EVOO consumption, but not MUFAs alone, was associated with a reduced risk of all-cause mortality, CVD events, and stroke which indicates that the minor constituents in EVOO could be also responsible for its health benefits. Furthermore, a study conducted by Guasch-Ferré et al. have suggested that 10 g/day of EVOO intake were associated with a 10% reduction in the risk of cardiovascular problems [92]. Another study suggests that EVOO is associated with lower CVD rates in people adhering to MD; in general, those consuming >0.5 tablespoon/day of olive oil had a 14% lower risk of CVD and an 18% lower risk of coronary heart disease [93].

4.2.2 Cancers

ROS have been implicated in the etiology of many cancer types; therefore a "control" over them would significantly reduce the risk of developing certain cancers [22]. There is a strong evidence that using EVOO as a main source of fats in someone's diet can supress certain cancer types, such as breast [4, 7, 82, 90, 94], colorectal [3, 7, 95], and prostate cancer [3, 94] due to its anti-oxidant and anti-inflammatory effects. The cancer-preventing mechanisms of EVOO, in general, are less known. It has been hypothesized that the anti-cancer actions of EVOO may relate to the ability of its MUFA-oleic acid to specifically regulate cancer-related oncogenes. In fact, oncologists have discovered that MUFA and PUFA suppress the over-expression of an oncogene HER2, which is critical to the etiology, invasion, progression, and metastasis especially of human mammary carcinoma [96, 97]. However, there is a growing interest to identify the role of phenolics from EVOO in carcinogenesis, as well. They can exert an inhibitory action on cancers, acting as blocking and/or suppressive agents at several stages of cancer progression [4, 8].

In vitro studies have reported that some phenolic compounds isolated from EVOO have anticancer activity against different types of cancers. Despite the different molecular mechanisms of the anticancer activities of EVOO phenolic compounds, it can be summarized that most of them inhibit oncogenic factors, including mutations, catalytic activities of predicted metabolic and epigenetic targets and interactions affecting DNA methylation. For instance, the inhibition of prostate cancer by hydroxytyrosol was found to be mediated by inhibition of cell proliferation, adhesion, migration, and invasion. Oleuropein has also demonstrated a chemo-preventive role in the proliferation of breast cancer cells by inhibiting estrogen-dependent signals. Oleocanthal and oleacein have reduced the viability and migration of non-melanoma skin cancer cells, while hydroxytyrosol showed no effect in this cancer type. Moreover, the metabolites produced by the degradation of EVOO phenolic compounds by gut microbiota may have a chemo-preventive effects on colorectal cancer, which is the second most common cancer-related death worldwide [45, 46, 98–102].

Another *in vitro* study of anticancer and chemo-preventive potential of EVOO's tyrosol, hydroxytyrosol and secoiridoid derivatives (oleocanthal and oleacein) on cutaneous non-melanoma skin cancer models have demonstrated that phenolic EVOO's compounds can block molecular steps that occur after the initial UV radiation exposure and before or during tumor development. Another trial investigated whether hydroxytyrosol improves the antitumor response of women with breast cancer undergoing neoadjuvant chemotherapy, influencing plasma levels of molecules involved in cell proliferation, apoptosis, and metastasis (e.g., tissue inhibitor of metalloproteinases, TIMP-1); data showed that in women receiving a dietary supplement with 15 mg/day of hydroxytyrosol combined with a specific chemotherapy treatment, the plasma levels of TIMP-1 decreased [46, 103–105]. Furthermore, EVOO induces molecular changes in tumors, such as in the composition of cell membranes, activity of signaling proteins and gene expression; all these modifications could cause lower proliferation, higher apoptosis and lower DNA damage. For example, there is a beneficial evidence of EVOO on breast cancer risk; consumption of EVOO in moderate quantities and throughout the lifetime appears to be a healthy choice and may favorably influence the breast cancer risk [4]. In another trial, the PREDIMED trial [91], part of the participants were subjected to a traditional MD supplemented with EVOO and compared to the participants subjected to the control low-fat diet. Besides the fact that the main outcome in the trial was the incidence of CVD, the incidence of breast cancer in women was included as a secondary outcome. Among 4152 women included in the analysis, women subjected to the MD supplemented with EVOO group exhibited a 68% lowest risk of incident breast cancer.

4.2.3 Other diseases

Metabolic syndrome (MS) is characterized by a cluster of interrelated markers including obesity, hyperglycaemia, dyslipidaemia and hypertension [4, 21, 42, 106–110]. EVOO's phenolic compounds have been related to the prevention or inhibition of MS-related diseases. An *in vitro* assay has demonstrated that oleacein acts as an inhibitor of a central epigenetic regulator of metabolic reprogramming in diseases associated with obesity, neurological disorders, and cancer. Inhibitory effects were also found against enzymes related to hyperglycemia associated with hypertension [111]. Positive results were found in a two-year study on subjects with MS, showing a reduction in blood pressure values in women with moderate hypertension supplemented with EVOO versus the control low-fat diet group [112].

Positive modulation of *gut microbiota* as an approach for management of diseases has attracted a lot of scientific attention recently. High intake of phenolic compounds from dietary source, such as EVOO, appears to regulate the CVD and other risk factors, through the modulation of gut microbial populations, their activities and diversity. This is due to the fact that many plants' phenolic compounds are not totally absorbed in the mouth and gastrointestinal track and become available for microbiota utilization as an energy source. This is the case with EVOO's phenolic compounds – reaching the gut and being consumed by its microbiota. This is very important since the gut microbiota, as a key factor in driving metabolic activities, is involved in the regulation of host immunity. Thus, supporting healthy gut microbiota can help boosting the overall immunity [24, 27, 82, 113–115].

Diabetes type II (T2D) could be prevented by using EVOO due to its rich phenolic profile [11, 22, 36, 42]. The specific components of EVOO are considered as novel candidates for improving the glycaemic profile in patients with diabetes mellitus. In an interventional study, it has been shown that oleuropein lowers postprandial glycaemia by reduction of Nox2 activity in healthy subjects. Another evidence is improved glucose metabolism and reduced body weight in the PREDIMED study, thus, preventing T2D in 80 cases of new-onset diabetes cases which were subjected to the MD and EVOO group of participants. More particularly, the positive effects of polyphenols on the metabolic control and the production of specific pro−/anti-inflammatory adipokines in overweight patients with T2D have been confirmed. EVOO consumption significantly reduced fasting plasma glucose and HbA1c levels, as well as body mass index and body weight. Therefore, daily consumption of polyphenol-rich EVOO might improve the metabolic control and circulating inflammatory adipokines profile in overweight T2D patients [110].

Reports on reduction in body weight by using EVOO in the diet have also been published [90, 116, 117]. *Chronic obesity* is a situation of chronic systemic inflammation and can contribute to the development and severity of asthmatic and probably allergic diseases [11, 42]. There is a limited evidence for the EVOO effects on the body weight regarding the reduction of fat mass with consequent increases of muscle mass. However, a trial evaluating the effects of hydroxytyrosol has showed that it can modulate the adipocyte gene expression profile through mechanisms involving a reduction of oxidative stress and NF-kB inhibition and may blunt macrophage recruitment, preventing the deregulation of pathways involved in the obesity-related diseases [11].

Antimicrobial activities have been found in olive extracts; EVOO components, like tocopherols, carotenoids, have been shown to reduce the growth of foodborne pathogens and stimulation of growth of probiotic microorganisms such as *L. Acidophilus* and *B. Bifidum* [17]. Also, EVOO phenols have been demonstrated to inhibit *in vivo* or delay the growth of bacteria, such as salmonella, cholera, pseudomonas, staphylococcus, fungi, viruses, and parasites. Furthermore, hydroxytyrosol has showed to be highly toxic to *Pseudomonas syringae* pv *savastanoi* and *Corynebacterium Michiganense,* while oleuropein can completely inhibit the growth of *Escherichia coli, Klebsiella pneuomoniae,* and *Bacillus cereus*. Also, the phenolics exert *in vitro* strong bactericidal activity against eight strains of *Helicobacter pylori*, which are linked to a majority of peptic ulcers and certain types of gastric cancers [8]. Therefore, there is a strong evidence of the antimicrobic effect of EVOO's phenols that successfully destroy colonies of microorganisms which may cause respiratory tract, intestinal, and genital tract infections [8, 90].

Autoimmune diseases and immune-mediated inflammatory diseases [7, 90, 118, 119] like inflammatory bowel disease, rheumatoid arthritis, systemic lupus erythematosus, sclerosis, psoriasis, etc., can be prevented by the favorable fat profile of EVOO. But, the beneficial effects of EVOO have been attributed, besides to the MUFA content, to the presence of phenolic compounds as well due to their antioxidant, anti-inflammatory and immunomodulatory properties [118]. There are many studies that support the beneficial role of EVOO in certain inflammatory diseases; for example, the positive effects of dietary EVOO on type II collagen-induced arthritis have been reported [4, 118, 119].

Skin treatments with EVOO after burning have resulted in better wound healing; or treatment with oleuropein and olive leaf extract after UV exposure significantly inhibited increase in skin thickness, reduction in skin elasticity, skin carcinogenesis and tumor growth. Furthermore, EVOO polyphenols exerted anti-inflammatory effects on human keratinocytes suppressing the key epidermal cytokines, which have been found as a source of skin inflammation [4, 90, 120].

EVOO has been proposed to promote *healthy aging*, as it is being able to virtually modulate all the features of the aging process, because of its favorable MUFA content and its minor bioactive compounds. Polyphenols are able to modulate abnormal cellular signaling induced by pro-inflammatory stimuli and oxidative stress. In particular, oleocanthal and oleacein are scientifically proved that can activate healthy aging-promoting cytoprotective pathways and suppress the oxidative stress in mammalian cells [4, 21, 47].

5. Summary

Generation of oxidative species in the body, such as free radicals, ROS and RNS, result in oxidative stress that has been implicated in the pathogenesis of many chronic degenerative diseases and aging processes. In particular, the oxidative stress leads to structural and functional damage to the main biomolecules, nucleic acids, lipids, and proteins, which then result in development of many diseases, such as cancer, metabolic disorders, cardiovascular dysfunctions, inflammatory disorders, neurological degeneration, etc. Antioxidants can counteract the effects of the free radicals and other oxidative species, thus reducing the oxidative stress in the body. An imbalance between the production of oxidative species from one side, and the availability of antioxidants from the other side, can be detrimental for the human health. Therefore, the level and the diversity of antioxidants in the body are crucial for counteracting and neutralizing the oxidative stress. *de novo* antioxidant production in the body is limited; therefore, dietary antioxidants, such as polyphenols, carotenoids, tocopherols, tocotrienols, and others, are preferred as they are naturally bioavailable compounds easily absorbed by the human body and ready to counteract the oxidative stress. In this context, EVOO is a great source of variety of powerful and bioavailable components, such as phenolic compounds hydroxytyrosol, oleocanthal and oleuropein. The health properties of EVOO have been directly related to these phenolic compounds. *In vitro* and *in vivo* studies involving humans and animals have demonstrated that EVOO's unsaturated fatty acids and phenolic compounds have remarkable and beneficial effects due to their pronounced anti-oxidant and anti-inflammatory activity, all acting in a synergistic way. A large body of scientific evidence has shown EVOO's effects on different pathologies, such as different types of cancers, CVDs, neurodegenerative and digestive disorders, MS and gut-related disorders.

In summary, EVOO - the hallmark of the Mediterranean diet - has been proven to have the following health and well-being benefits (**Figure 4**):

- Reduced risk of cancers,
- Improved endothelial function,
- Reduced blood pressure,
- Antithrombotic effects,
- Reduced total cholesterol level,
- Better glycemic control,
- Reduced metabolic-related diseases,
- Antimicrobial activity,
- Reduced inflammation,
- Yields rich gut microbiota,
- Supports healthy aging, etc.

Taking into account the above benefits of EVOO, it is obvious that EVOO should be used in abundance in the daily diet, as it is the case with the people adhering to the remarkable Mediterranean Diet, who are experiencing lower risks of certain diseases that could be avoided via diet. EVOO - the "*Blessed nutrient*" and the "*Elixir of youth and health*" known from the ancient times - is indeed the proof of the Hippocrates's "*Let food be thy medicine and medicine be thy food*".

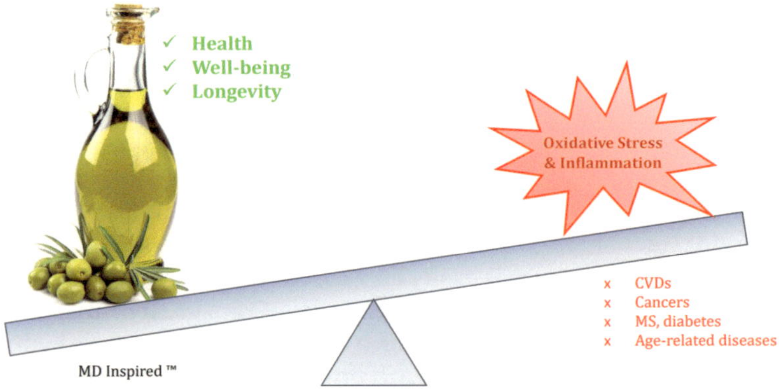

Figure 4.
EVOO's antioxidant and anti-inflammatory compounds efficiently neutralize the oxidative stress and fight the inflammation.

Acknowledgements

The authors sincerely acknowledge the support of CSI: Create.Solve. Innovate. LLC and its concept MD Inspired™ for promoting the Mediterranean lifestyle outside the Mediterranean region.

Conflict of interest

The authors declare no conflict of interest.

Author details

Anka Trajkovska Petkoska* and Anita Trajkovska-Broach
CSI: Create. Solve. Innovate. LLC, Blacksburg, VA, USA

*Address all correspondence to: anka@createsolveinnovate.com

References

[1] Moro E. The Mediterranean Diet from Ancel Keys to the UNESCO Cultural Heritage. A Pattern of Sustainable Development between Myth and Reality. Procedia - Social and Behavioral Sciences. 2016; 223: 655-661. doi: 10.1016/j.sbspro.2016.05.380

[2] Foscolou A, Critselis E, Panagiotakos D. Olive oil consumption and human health: A narrative review. Maturitas. 2018; 118: 60-66. doi.org/10.1016/j.maturitas.2018.10.013

[3] Salazar DM, López-Cortés I, Salazar-García DC. Olive oil: composition and health benefits. In: Fritjof T, Henning R, editors. Olive oil - sensory characteristics, composition and importance in human health. Nova Science Publishers Inc; 2017. ISBN: 978-1-53612-563-4.

[4] Ben-Noun L (Nun). Health benefits of olive oil. B.N. Publication House (Israel); 2018.

[5] Franconi F, Campesi I, Romani A. Is Extra Virgin Olive Oil an Ally for Women's and Men's Cardiovascular Health? Cardiovascular Terapeutics. 2020; 6719301. doi.org/10.1155/2020/6719301

[6] Cicerale S, Lucas LJ, Keast RSJ. Antimicrobial, antioxidant and anti-inflammatory phenolic activities in extra virgin olive oil. Current Opinion in Biotechnology. 2012; 23:129-135. DOI 10.1016/j.copbio.2011.09.006

[7] Gaforio JJ, Visioli F, Alarcón-de-la-Lastra C, Castañer O, Delgado-Rodríguez M, Fitó M, Hernández AF, Huertas JR, Martínez-González MA, Menendez JA, de la Osada J, Papadaki A, Parrón T, Pereira JE, Rosillo MA, Sánchez-Quesada C, Schwingshackl L, Toledo E, Tsatsakis AM. Virgin Olive Oil and Health: Summary of the III International Conference on Virgin Olive Oil and Health Consensus Report, JAEN (Spain) 2018. Nutrients. 2019; 11: 2039. doi:10.3390/nu11092039

[8] Boskou D, editor. Olive oil - minor constituents and health. Taylor & Francis Group, LLC; 2009.

[9] Visioli F, Franco M, Toledo E, Luchsinger J, Willett WC, Hu FB, Martinez-Gonzalez MA. Olive oil and prevention of chronic diseases: Summary of an International conference. Nutrition, Metabolism & Cardiovascular Diseases. 2018; 28: 649e656. doi.org/10.1016/j.numecd.2018.04.004

[10] Nocella C, Cammisotto V, Fianchini L, D'Amico A, Novo M, Castellani V, Stefanini L, Violi F, Carnevale R. Extra Virgin Olive Oil and Cardiovascular Diseases: Benefits for Human Health, Endocrine, Metabolic & Immune Disorders - Drug Targets. 2018; 18: 4-13. DOI:10.2174/1871530317666171114121533

[11] Mazzocchi A, Leone L, Agostoni C, Pali-Schöll I.The Secrets of the Mediterranean Diet. Does [Only] Olive Oil Matter? Nutrients. 2019; 11: 2941. doi:10.3390/nu11122941

[12] Trajkovska Petkoska A, Broach TA. Mediterranean Way of Living as an Optimal Lifestyle and a Dietary Pattern for Healthy Gut and Strong Immunity, EC Nutrition. 2021; 16.1: 141-171.

[13] Guasch-Ferré M, Liu G, Li Y, Sampson L, Manson JE, Salas-Salvadó J, Martínez-González MA, Stampfer MJ, Willett WC, Sun Q, Hu FB. Olive Oil Consumption and Cardiovascular Risk in U.S. Adults. Journal of the American College of Cardiology. 2020. doi.org/10.1016/j.jacc.2020.02.036

[14] Tejada S, Pinya S, del Mar Bibiloni M, Tur JA, Pons A,

Sureda A. Cardioprotective Effects of the Polyphenol Hydroxytyrosol from Olive Oil. Current Drug Targets. 2017; 18: 1477-1486. DOI:10.2174/1389450117 66616100515065

[15] Giacosa A. Cancer prevention in Europe: the Mediterranean diet as a protective choice. European Journal of Cancer Prevention. 2013; 22:90-95. DOI: 10.1097/CEJ.0b013e328354d2d716.

[16] Rodríguez-López P, Lozano-Sanchez J, Borrás-Linares I, Emanuelli T, Menéndez JA, Segura-Carretero A. Structure–Biological Activity Relationships of Extra-Virgin Olive Oil Phenolic Compounds: Health Properties and Bioavailability. Antioxidants. 2020; 9: 685. doi:10.3390/antiox9080685

[17] Mehmood A, Usman M, Patil P, Zhao L, Wang C. A review on management of cardiovascular diseases by olive polyphenols. Food Sci Nutr. 2020; 8:4639-4655. DOI: 10.1002/fsn3.1668

[18] Lozano-Castellón J, López-Yerena A, Rinaldi de Alvarenga JF, del Castillo-Alba JR, Vallverdú-Queralt A, Escribano-Ferrer E, Lamuela-Raventós RM. Health-promoting properties of oleocanthal and oleacein: Two secoiridoids from extra-virgin olive oil. Critical Reviews in Food Science and Nutrition. 2020; 60(15):2532-2548. DOI:10.1080/10408398.2019.1650715

[19] Rocchetti G, Senizza B, Giuberti G, Montesano D, Trevisan M, Lucini L. Metabolomic Study to Evaluate the Transformations of Extra-Virgin Olive Oil's Antioxidant Phytochemicals During In Vitro Gastrointestinal Digestion. Antioxidants. 2020; 9: 302. doi:10.3390/antiox9040302

[20] Atlante A, Amadoro G, Bobba A, Latina V. Functional Foods: An Approach to Modulate Molecular Mechanisms of Alzheimer's Disease. Cells. 2020; 9: 2347. doi:10.3390/cells9112347

[21] Serreli G, Deiana M. Extra Virgin Olive Oil Polyphenols: Modulation of Cellular Pathways Related to Oxidant Species and Inflammation in Aging. Cells. 2020; 9(2): 478. doi:10.3390/cells9020478

[22] Gorzynik-Debicka M, Przychodzen P, Cappello F, Kuban-Jankowska A, Gammazza AM, Knap N, Wozniak M, Gorska-Ponikowska M. Potential Health Benefits of Olive Oil and Plant Polyphenols. Int. J. Mol. Sci. 2018; 19: 547. doi:10.3390/ijms19030686

[23] Romani A, Ieri F, Urciuoli S, Noce A, Marrone G, Nediani C, Bernini R. Health Effects of Phenolic Compounds Found in Extra-Virgin Olive Oil, By-Products, and Leaf of *Olea europaea* L. Nutrients. 2019; 11(8): 1776. doi: 10.3390/nu11081776

[24] Jimenez-Lopez C, Carpena M, Lourenço-Lopes C, Gallardo-Gomez M, Lorenzo JM, Barba FJ, Prieto MA, Simal-Gandara J. Bioactive Compounds and Quality of Extra Virgin Olive Oil. Foods. 2020; 9: 1014. doi:10.3390/foods9081014

[25] Presti G, Guarrasi V, Gulotta E, Provenzano F, Provenzano A, Giuliano S, Monfreda M, Mangione MR, Passantino R, San Biagio PL, Costa MA, Giacomazza D. Bioactive compounds from extra virgin olive oils: Correlation between phenolic content and oxidative stress cell protection. Biophysical Chemistry. 2017; 230: 109-116. doi.org/10.1016/j.bpc.2017.09.002

[26] Guasch-Ferré, M., Hu, F.B., Martínez-González, M.A. *et al.* Olive oil intake and risk of cardiovascular disease and mortality in the PREDIMED Study. BMC Med. 2014; 12: 78. doi.org/10.1186/1741-7015-12-78

[27] Marcelino G, Hiane PA, de Cássia Freitas K, Santana LF, Pott A, Donadon JR, de Cássia Avellaneda Guimarães R. Effects of Olive Oil and Its

Minor Components on Cardiovascular Diseases, Inflammation, and Gut Microbiota. Nutrients. 2019; 11: 1826. doi:10.3390/nu11081826

[28] Lanza B, Ninfali P. Antioxidants in Extra Virgin Olive Oil and Table Olives: Connections between Agriculture and Processing for Health Choices. Antioxidants. 2020; 9: 41. doi:10.3390/antiox9010041

[29] Carluccio MA, Calabriso N, Egeria Scoditti, Massaro M, De Caterina R. Mediterranean Diet Polyphenols. In The Mediterranean Diet. Elsevier; 2015.

[30] Hoffman R, Gerber M. Food Processing and the Mediterranean Diet. Nutrients. 2015; 7: 7925-7964. doi:10.3390/nu7095371

[31] Fernandes J, Fialho M, Santos R, Peixoto-Placido C, Madeira T, Sousa-Santos N, Virgolino A, Santos O, Vaz Carneiro A. Is olive oil good for you? A systematic review and meta-analysis on anti-inflammatory benefits from regular dietary intake. Nutrition. 2020; 69: 110559. doi.org/10.1016/j.nut.2019.110559

[32] Francisco V, Ruiz-Fernández C, Lahera V, Lago F, Pino J, Skaltsounis L, González-Gay MA, Mobasheri A, Gómez R, Scotece M, Gualillo O. Natural Molecules for Healthy Lifestyles: Oleocanthal from Extra Virgin Olive Oil. J. Agric. Food Chem. 2019; 67: 3845-3853. DOI: 10.1021/acs.jafc.8b06723

[33] Klimova B, Novotný M, Kuca K, Valis M. Effect Of An Extra-Virgin Olive Oil Intake On The Delay Of Cognitive Decline: Role Of Secoiridoid Oleuropein? Neuropsychiatric Disease and Treatment. 2019;15: 3033-3040.

[34] Hidalgo-Mora JJ, Cortés-Sierra L, García-Pérez MA, Tarín JJ, Cano A. Diet to Reduce the Metabolic Syndrome Associated with Menopause. The Logic for Olive Oil. Nutrients. 2020; 12: 3184. doi:10.3390/nu12103184

[35] Perez-Jimenez F, Ruano J, Perez-Martinez P, Lopez-Segura F, Lopez-Miranda J. The influence of olive oil on human health: not a question of fat alone. Mol. Nutr. Food Res. 2007; 51: 1199-1208. DOI 10.1002/mnfr.200600273

[36] Lopez-Miranda J et al. Olive oil and health: Summary of the II international conference on olive oil and health consensus report, Jae'n and Co'rdoba (Spain) 2008. Nutrition, Metabolism & Cardiovascular Diseases. 2010; 20: 284e294. doi:10.1016/j.numecd.2009.12.007

[37] Biswas SK. Does the Interdependence between Oxidative Stress and Inflammation Explain the Antioxidant Paradox? Oxidative Medicine and Cellular Longevity. 2016; 5698931. doi.org/10.1155/2016/5698931

[38] Chatterjee S. Oxidative Stress, Inflammation, and Disease. In: Oxidative Stress and Biomaterials. Elsevier Inc. 2016. doi.org/10.1016/B978-0-12-803269-5.00002-4

[39] Hussain T, Tan B, Yin Y, Blachier F, Tossou MCB, Rahu N. Oxidative Stress and Inflammation: What Polyphenols Can Do for Us? Oxidative Medicine and Cellular Longevity. 2016; 7432797. http://dx.doi.org/10.1155/2016/7432797

[40] Simpson DSA, Oliver PL. ROS Generation in Microglia: Understanding Oxidative Stress and Inflammation in Neurodegenerative Disease. Antioxidants. 2020; 9: 743. doi:10.3390/antiox9080743

[41] Visioli F, Poli A, Galli C. Oleocanthal, a Phenolic Derived from Virgin Olive Oil: A Review of the Beneficial Effects on Inflammatory Disease. Antioxidant and Other Biological Activities of Phenols from Olives and Olive Oil. Medicinal Research Reviews. 2002; 22(1): 65-75. DOI 10.1002/med.1028

[42] De Santis S, Cariello M, Piccinin E, Sabbà C, Moschetta A. Extra Virgin Olive Oil: Lesson from Nutrigenomics. Nutrients. 2019; 11: 2085. doi:10.3390/nu11092085

[43] Gambino CM, Accardi G, Aiello A, Candore G, Dara-Guccione G, Mirisola M, Procopio A, Taormina G, Caruso C. Effect of Extra Virgin Olive Oil and Table Olives on the ImmuneInflammatory Responses: Potential Clinical Applications. Endocrine, Metabolic & Immune Disorders - Drug Targets. 2018; 18: 14-22. DOI:10.2174/1871530317666171114113822

[44] Criado-Navarro I, López-Bascón MA, Priego-Capote F. Evaluating the variability in the phenolic concentration of extra virgin olive oil according to the Commission Regulation (EU) 432/2012 health claim. J. Agric. Food Chem. 2020. DOI: 10.1021/acs.jafc.0c02380

[45] Baci, D.; Gallazzi, M.; Cascini, C.; Tramacere, M.; De Stefano, D.; Bruno, A.; Noonan, D.M.; Albini, A. Downregulation of pro-inflammatory and pro-angiogenic pathways in prostate cancer cells by a polyphenol-rich extract from olive mill wastewater. Int. J. Mol. Sci. 2019; 20: 307. doi: 10.3390/ijms20020307

[46] Polini B, Digiacomo M, Carpi S, Bertini S, Gado F, Saccomanni G, Macchia M, Nieri P, Manera C, Fogli S. Oleocanthal and oleacein contribute to the in vitro therapeutic potential of extra virgin oil-derived extracts in non-melanoma skin cancer. Toxicol. Vitr. 2018; 52: 243-250. doi: 10.1016/j.tiv.2018.06.021

[47] Nikou T, Liaki V, Stathopoulos P, Sklirou AD, Tsakiri EN, Jakschitz T, Bonn G, Trougakos IP, Halabalaki M, Skaltsounis LA. Comparison survey of EVOO polyphenols and exploration of healthy agingpromoting properties of oleocanthal and oleacein. Food and Chemical Toxicology. 2019; 125: 403-412. https://doi.org/10.1016/j.fct.2019.01.016

[48] Reboredo-Rodríguez P, Figueiredo-González M, González-Barreiro C, Simal-Gándara J, Salvador MD, Cancho-Grande B, Fregapane G. State of the art on functional virgin olive oils enriched with bioactive compounds and their properties. Int. J. Mol. Sci. 2017; 18(3): 668. DOI: 10.3390/ijms18030668

[49] Rosillo MA, Alcaraz MJ, Sánchez-Hidalgo M, Fernández-Bolaños JG, Alarcón-de-la-Lastra C, Ferrándiz ML. Anti-inflammatory and joint protective effects of extra-virgin olive-oil polyphenol extract in experimental arthritis. Journal of Nutritional Biochemistry. 2014; 25: 1275-1281. dx.doi.org/10.1016/j.jnutbio.2014.07.006

[50] Boskou D. Olive Fruit, Table Olives, and Olive Oil Bioactive Constituents.

[51] Fernandes J, Fialho M, Santos R, Peixoto-Plácido C, Madeira T, Sousa-Santos N, Virgolino A, Santos O, Vaz Carneiro A. Is olive oil good for you? A systematic review and meta-analysis on anti-inflammatory benefits from regular dietary intake. Nutrition. 2020; 69:110559. doi: 10.1016/j.nut.2019.110559

[52] D'Angelo C, Franceschelli S, Quiles JL, Speranza L. Wide Biological Role of Hydroxytyrosol: Possible Therapeutic and Preventive Properties in Cardiovascular Diseases. Cells. 2020; 9: 1932. doi:10.3390/cells9091932

[53] Bertelli M, Kiani AK, Paolacci S, Manara E, Kurti D, Dhuli K, Bushati V, Miertus J, Pangallo D, Baglivo M, Beccari T, Michelini S. Hydroxytyrosol: A natural compound with promising pharmacological activities. Journal of Biotechnology. 2020; 309: 29-33. doi.org/10.1016/j.jbiotec.2019.12.016

[54] Robles-Almazan M, Pulido-Moran M, Moreno-Fernandez J, Ramirez-Tortosa C, Rodriguez-Garcia C, Quiles JL, Ramirez-Tortosa MC.

Hydroxytyrosol: Bioavailability, toxicity, and clinical applications. Food Research International. 2018; 105: 654-667. doi.org/10.1016/j.foodres.2017.11.053

[55] Granados-Principal C, José L Quiles JL, Ramirez-Tortosa CL, Pedro Sanchez-Rovira P, Ramirez-Tortosa MC. Hydroxytyrosol: from laboratory investigations to future clinical trials. Nutrition Reviews. 2010; 68(4): 191-206. doi:10.1111/j.1753-4887.2010.00278.x

[56] Calabriso N, Massaro M, Scoditti E, D'Amore S, Gnoni A, Pellegrino M, Storelli C, De Caterina R, Palasciano G, Carluccio MA. Extra virgin olive oil rich in polyphenols modulates VEGF-induced angiogenic responses by preventing NADPH oxidase activity and expression. J. Nutr. Biochem. 2016; 28: 19-29. doi: 10.1016/j.jnutbio.2015.09.026

[57] Palmieri D, Aliakbarian B, Casazza AA, Ferrari N, Spinella G, Pane B, Cafueri G, Perego P, Palombo D. Effects of polyphenol extract from olive pomace on anoxia-induced endothelial dysfunction. Microvasc. Res. 2012; 83: 281-289. DOI: 10.1016/j.mvr.2012.02.010

[58] Zrelli H, Matsuoka M, Kitazaki S, Zarrouk M, Miyazaki H. Hydroxytyrosol reduces intracellular reactive oxygen species levels in vascular endothelial cells by upregulating catalase expression through the AMPK-FOXO3a pathway. Eur. J. Pharmacol. 2011; 660: 275-282. doi: 10.1016/j.ejphar.2011.03.045

[59] Zhu L, Liu Z, Feng Z, Hao J, Shen W, Li X, Sun L, Sharman E, Wang Y, Wertz K, Weber P, Shi X, Liu J. Hydroxytyrosol protects against oxidative damage by simultaneous activation of mitochondrial biogenesis and phase II detoxifying enzyme systems in retinal pigment epithelial cells. J. Nutr. Biochem. 2010; 21: 1089-1098. DOI: 10.1016/j.jnutbio.2009.09.006

[60] Sánchez-Fidalgo S, Villegas I, Aparicio-Soto M, Cárdeno A, Rosillo MÁ, González-Benjumea A, Marset A, López Ó, Maya I, Fernández-Bolaños JG, Alarcón de la Lastra C. Effects of dietary virgin olive oil polyphenols: Hydroxytyrosyl acetate and 3, 4-dihydroxyphenylglycol on DSS-induced acute colitis in mice. J. Nutr. Biochem. 2015; 26: 513-520. DOI: 10.1016/j.jnutbio.2014.12.001

[61] Sánchez-Fidalgo S, Sánchez De Ibargüen L, Cárdeno A, Alarcón De La Lastra C. Influence of extra virgin olive oil diet enriched with hydroxytyrosol in a chronic DSS colitis model. Eur. J. Nutr. 2012; 51: 497-506. doi: 10.1007/s00394-011-0235-y

[62] Nazzaro F, Fratianni F, Cozzolino R, Martignetti A, Malorni L, De Feo V, Cruz AG, d'Acierno A. Antibacterial Activity of Three Extra Virgin Olive Oils of the Campania Region, Southern Italy, Related to Their Polyphenol Content and Composition. Microorganisms. 2019; 7: 321. doi:10.3390/microorganisms7090321

[63] Patti AM, Carruba G, Cicero AFG, Banach M, Nikolic D, Giglio RV, Terranova A, Soresi M, Giannitrapani L, Montalto G, Stoian AP, Banerjee Y, Rizvi AA, Toth PP, Rizzo M. Daily Use of Extra Virgin Olive Oil with High Oleocanthal Concentration Reduced Body Weight, Waist Circumference, Alanine Transaminase, Inflammatory Cytokines and Hepatic Steatosis in Subjects with the Metabolic Syndrome: A 2-Month Intervention Study. Metabolites. 2020; 10: 392. doi:10.3390/metabo10100392

[64] Scotece M. Conde J, Abella V, Lopez V, Pino J, Lago F, Smith AB 3rd, Gómez-Reino JJ, Gualillo O. New drugs from ancient natural foods. Oleocanthal, the natural occurring spicy compound of olive oil: a brief history. Drug Discov Today. 2015; 20(4):406-10. doi.org/10.1016/j.drudis.2014.10.017

[65] Papanikolaou C, Eleni Melliou E, Magiatis P. Olive Oil Phenols. doi.org/10.5772/intechopen.81394

[66] Patti AM, Carruba G, Cicero AFG, Banach M, Nikolic D, Giglio RV, Terranova A, Soresi M, Giannitrapani L, Montalto G, Stoian AP, Banerjee Y, Rizvi AA, Toth PP, Rizzo M. Daily Use of Extra Virgin Olive Oil with High Oleocanthal Concentration Reduced Body Weight, Waist Circumference, Alanine Transaminase, Inflammatory Cytokines and Hepatic Steatosis in Subjects with the Metabolic Syndrome: A 2-Month Intervention Study. Metabolites. 2020; 10: 392. doi:10.3390/metabo10100392

[67] Beauchamp GK, Keast RSJ, Morel D, Lin J, Pika J, Han Q, Lee CH, Smith AB, Breslin PAS. Ibuprofen-like activity in extra-virgin olive oil. Nature. 2005; 437. doi:10.1038/437045a

[68] Cicerale S, Conlan XA, Barnett NW, Andrew J, Sinclair AJ, Keast RSJ. Influence of Heat on Biological Activity and Concentration of Oleocanthalsa Natural Anti-inflammatory Agent in Virgin Olive Oil. J. Agric. Food Chem. 2009; 57: 1326-1330. doi: 10.1021/jf803154w

[69] Aparicio-Soto M, Sánchez-Fidalgo S, González-Benjumea A, Maya I, Fernández-Bolaños JG, Alarcón-de-la-Lastra C. Naturally occurring hydroxytyrosol derivatives: Hydroxytyrosyl acetate and 3,4-dihydroxyphenylglycol modulate inflammatory response in murine peritoneal macrophages. Potential utility as new dietary supplements. J. Agric. Food Chem. 2015; 63: 836-846. DOI: 10.1021/jf503357s

[70] Sirianni R, Chimento A, de Luca A, Casaburi I, Rizza P, Onofrio A, Iacopetta D, Puoci F, Andò S, Maggiolini M, Pezzi V. Oleuropein and hydroxytyrosol inhibit MCF-7 breast cancer cell proliferation interfering with ERK1/2 activation. Mol. Nutr. Food Res. 2010; 54: 833-840. doi: 10.1002/mnfr.200900111

[71] Bulotta S, Corradino R, Celano M, D'Agostino M, Maiuolo J, Oliverio M, Procopio A, Iannone M, Rotiroti D, Russo D. Antiproliferative and antioxidant effects on breast cancer cells of oleuropein and its semisynthetic peracetylated derivatives. Food Chem. 2011; 127(4): 1609-1614. Doi: 10.1016/j.foodchem.2011.02.025

[72] Bouallagui Z, Han J, Isoda H, Sayadi S. Hydroxytyrosol rich extract from olive leaves modulates cell cycle progression in MCF-7 human breast cancer cells. Food Chem. Toxicol. 2011; 49: 179-184. doi: 10.1016/j.fct.2010.10.014

[73] Paiva-Martins F, Fernandes J, Santos V, Silva L, Borges F, Rocha S, BeloL,Santos-SilvaA.Powerfulprotective role of 3,4-dihydroxyphenylethanol-elenolic acid dialdehyde against erythrocyte oxidative-Lnduced hemolysis. J. Agric. Food Chem. 2010; 58: 135-140. doi: 10.1021/jf9031052

[74] Lucas L, Russell A, Keast R. Molecular Mechanisms of Inflammation. Anti-Inflammatory Benefits of Virgin Olive Oil and the Phenolic Compound Oleocanthal. Curr. Pharm. Des. 2011; 17: 754-768. doi: 10.2174/138161211795428911

[75] Cárdeno A, Sánchez-Hidalgo M, Aparicio-Soto M, Sánchez-Fidalgo S, Alarcón-De-La-Lastra C. Extra virgin olive oil polyphenolic extracts downregulate inflammatory responses in LPS-activatedmurine peritoneal macrophages suppressing NFkB and MAPK signalling pathways. Food Funct. 2014; 5: 1270-1277. doi: 10.1039/c4fo00014e

[76] Bartolomei M, Bollati C, Bellumori M, Cecchi L, Cruz-Chamorro I, Santos-Sánchez G, Ranaldi G, Ferruzza S, Sambuy Y,

Arnoldi A, Mulinacci N, Lammi C. Extra Virgin Olive Oil Phenolic Extract on Human Hepatic HepG2 and Intestinal Caco-2 Cells: Assessment of the Antioxidant Activity and Intestinal Trans-Epithelial Transport. *Antioxidants*. 2021; 10(1):118. doi.org/10.3390/antiox10010118

[77] Omar SH. Oleuropein in Olive and its Pharmacological Effects. Sci Pharm. 2010; 78: 133-154. doi:10.3797/scipharm.0912-18

[78] Nediani C, Ruzzolini J, Romani A, Calorini L. Oleuropein, a Bioactive Compound from *Olea europaea* L., as a Potential Preventive and Therapeutic Agent in Non-Communicable Diseases. Antioxidants. 2019; 8: 578. doi:10.3390/antiox8120578

[79] Almatroodi SA, Almatroudi A, Anwar S, Babiker AY, Khan AA, Alsahli MA, Rahmani AH. Antioxidant, anti-inflammatory and hepatoprotective effects of olive fruit pulp extract: invivo and invitro study. Journal of Taibah University for Science. 2020; 14:1, 1660-1670. DOI: 10.1080/16583655.2020.1848761

[80] Santini SJ, Porcu C, Tarantino G, Amicarelli F, Balsano C. Oleuropein overrides liver damage in steatotic mice. Journal of Functional Foods. 2020; 65: 103756. doi.org/10.1016/j.jff.2019.103756

[81] Yubero-Serrano EM, Lopez-Moreno J, Gomez-Delgado F, Lopez-Miranda J. Extra virgin olive oil: More than a healthy fat. Eur J Clin Nutr. 2019; 72: 8-17. doi.org/10.1038/s41430-018-0304-x

[82] Gavahian M, Khaneghah AM, Lorenzo JM, Munekata PES, Garcia-Mantrana I, Collado MC, Meléndez-Martínez AJ, Barba FJ. Health benefits of olive oil and its components: Impacts on gut microbiota antioxidant activities, and prevention of noncommunicable diseases. Trends in Food Science & Technology. 2019; 88: 220-227. doi.org/10.1016/j.tifs.2019.03.008

[83] D'Amore S, Vacca M, Cariello M, Graziano G, D'Orazio A, Salvia R, Sasso RC, Sabbà C, Palasciano G, Moschetta A. Genes and miRNA expression signatures in peripheral blood mononuclear cells in healthy subjects and patients with metabolic syndrome after acute intake of extra virgin olive oil. Biochim. Biophys. Acta - Mol. Cell Biol. Lipids. 2016; 1861: 1671-1680. doi: 10.1016/j.bbalip.2016.07.003

[84] Perrone MA, Gualtieri P, Gratteri S, Ali W, Sergi D, Muscoli S, Cammarano A, Bernardini S, Di Renzo L, Romeo F. Effects of postprandial hydroxytyrosol and derivates on oxidation of LDL, cardiometabolic state and gene expression: A nutrigenomic approach for cardiovascular prevention. J. Cardiovasc. Med. 2019; 20: 419-426. doi: 10.2459/JCM.0000000000000816

[85] Fernández-Castillejo S, García-Heredia AI, Solà R, Camps J, López de la Hazas MC, Farràs M, Pedret A, Catalán Ú, Rubió L, Motilva MJ, Castañer O, Covas MI, Valls RM. Phenol-enriched olive oils modify paraoxonase-related variables: A randomized, crossover, controlled trial. Mol. Nutr. Food Res. 2017; 61(10). https://doi.org/10.1002/mnfr.201600932

[86] Pedret A, Fernández-Castillejo S, Valls RM, Catalán Ú, Rubió L, Romeu M, Macià A, López de las Hazas MC, Farràs M, Giralt M, Mosele JI, Martín-Peláez S, Remaley AT, Covas MI, Fitó M, Motilva MJ, Solà R. Cardiovascular Benefits of Phenol-Enriched Virgin Olive Oils: New Insights from the Virgin Olive Oil and HDL Functionality (VOHF) Study. Mol. Nutr. Food Res. 2018; 62: e1800456. doi:10.1002/mnfr.201800456

[87] Covas MI, Fitó M, de la Torre R. Minor Bioactive Olive Oil Components

and Health: Key Data for Their Role in Providing Health Benefits in Humans.

[88] Mitjavila MT, Fandos M, Salas-Salvadó J, Covas MI, Borrego S, Estruch R Lamuela-Raventós R, Corella D, Martínez-Gonzalez MÁ, Sánchez JM, Bulló M, Fitó M, Tormos C, Cerdá C, Casillas R, Moreno JJ, Iradi A, Zaragoza C, Chaves J, Sáez GT. The Mediterranean diet improves the systemic lipid and DNA oxidative damage in metabolic syndrome individuals. A randomized, controlled, trial. Clin. Nutr. 2013; 32: 172-178. doi: 10.1016/j.clnu.2012.08.002

[89] Bogani P, Galli C, Villa M, Visioli F. Postprandial anti-inflammatory and antioxidant effects of extra virgin olive oil. Atherosclerosis. 2007; 190: 181-186. doi: 10.1016/j.atherosclerosis.2006.01.011

[90] Fistonić I, Šitum M, Bulat V, Harapin M, Fistonić N, Verbanac D. Olive oil biophenols and women's health. Med. Glas. 2012; 9(1):1-9.

[91] Estruch R, et al., for the PREDIMED Study Investigators. Primary Prevention of Cardiovascular Disease with a Mediterranean Diet. N Engl J Med. 2013; 368: 1279-1290. DOI: 10.1056/NEJMoa1200303

[92] Guasch-Ferré M, Hu FB, Martínez-González MA, Fitó M, Bulló M, Estruch R, Ros E, Corella D, Recondo J, Gómez-Gracia E, et al. Olive oil intake and risk of cardiovascular disease and mortality in the PREDIMED Study. BMC Med. 2014; 12: 78. doi.org/10.1186/1741-7015-12-78

[93] Estruch R, Lamuela-Raventós RM, Ros E. The Bitter Taste of Extra Virgin Olive Oil for a Sweet Long Life. Journal of the american college of cardiology. 2020; 7 5(15): 1740-1742.

[94] Pauwels EKJ. The Protective Effect of the Mediterranean Diet: Focus on Cancer and Cardiovascular Risk. Med Princ Pract. 2011; 20: 103-111. DOI: 10.1159/000321197.

[95] Borzì AM, Biondi A, Basile F, Luca S, Vicari ESD, Vacante M. Olive Oil Effects on Colorectal Cancer. Nutrients. 2019; 11: 32. doi:10.3390/nu11010032

[96] Menendez JA, Vellon L, Colomer R, Lupu R. Oleic acid, the main monounsaturated fatty acid of olive oil, suppresses Her-2/neu (erbB-2) expression and synergistically enhances the growth inhibitory effects of trastuzumab (Herceptine) in breast cancer cells with Her-2/neu oncogene amplification. Annals of Oncology. 2005; 16: 359-371. doi:10.1093/annonc/mdi090

[97] Menéndez JA, Vázquez-Martín A, Ropero S, Colomer R, Lupu R. HER2 (erbB-2)-targeted effects of the omega-3 polyunsaturated fatty acid, alpha-linolenic acid (ALA; 18:3n-3), in breast cancer cells: the "fat features" of the "Mediterranean diet" as an "anti-HER2 cocktail". Clin Transl Oncol. 2006; 8(11):812-20. doi: 10.1007/s12094-006-0137-2. PMID: 17134970.

[98] Cuyàs E, Verdura S, Lozano-Sánchez J, Viciano I, Llorach-Parés L, Nonell-Canals A, Bosch-Barrera J, Brunet J, Segura-Carretero A, Sanchez-Martinez M, Encinar JA, Menendez JA. The extra virgin olive oil phenolic oleacein is a dual substrate-inhibitor of catechol-O-methyltransferase. Food Chem. Toxicol. 2019; 128: 35-45. doi.org/10.1016/j.fct.2019.03.049

[99] Cuyàs E, Castillo D, Llorach-Parés L, Lozano-Sánchez J, Verdura S, Nonell-Canals A, Brunet J, Bosch-Barrera J, Joven J, Valdés R, Sanchez-Martinez M, Segura-Carretero A, Menendez JA. Computational de-orphanization of the olive oil biophenol oleacein: Discovery of new metabolic and epigenetic targets. Food Chem. Toxicol. 2019; 131: 110529. DOI: 10.1016/j.fct.2019.05.037

[100] Cuyàs E, Gumuzio J, Lozano-Sánchez J, Carreras D, Verdura S, Llorach-Parés L, Sanchez-Martinez M, Selga E, Pérez GJ, Scornik FS, Brugada R, Bosch-Barrera J, Segura-Carretero A, Martin ÁG, Encinar JA, Menendez JA. Extra virgin olive oil contains a phenolic inhibitor of the histone demethylase LSD1/KDM1A. Nutrients. 2019; 11: 1656. doi: 10.3390/nu11071656

[101] Oliveras-Ferraros C, Fernández-Arroyo S, Vazquez-Martin, Lozano-Sánchez J, Cufí S, Joven J, Micol V, Fernández-Gutiérrez A, Segura-Carretero A, Menendez JA. Crude phenolic extracts from extra virgin olive oil circumvent de novo breast cancer resistance to HER1/HER2-targeting drugs by inducing GADD45-sensed cellular stress, G2/M arrest and hyperacetylation of Histone H3. Int. J. Oncol. 2011; 38, 1533-1547. DOI: 10.3892/ijo.2011.993

[102] Corominas-Faja B, Cuyàs E, Lozano-Sánchez J, Cufí S, Verdura S, Fernández-Arroyo S, Borrás-Linares I, Martin-Castillo B, Martin ÁG, Lupu R, Nonell-Canals A, Sanchez-Martinez M, Micol V, Joven J, Segura-Carretero A, Menendez JA. Extra-virgin olive oil contains a metabolo-epigenetic inhibitor of cancer stem cells. Carcinogenesis 2018; 39: 601-613. DOI: 10.1093/carcin/bgy023

[103] Ramirez-Tortosa C, Sanchez A, Perez-Ramirez C, Quiles JL, Robles-Almazan M, Pulido-Moran M, Sanchez-Rovira P, Ramirez-Tortosa M. Hydroxytyrosol Supplementation Modifies Plasma Levels of Tissue Inhibitor of Metallopeptidase 1 in Women with Breast Cancer. Antioxidants 2019; 8: 393. doi: 10.3390/antiox8090393

[104] Storniolo CE, Martínez-Hovelman N, Martínez-Huélamo M, Lamuela-Raventos RM, Moreno JJ. Extra Virgin Olive Oil Minor Compounds Modulate Mitogenic Action of Oleic Acid on Colon Cancer Cell Line. J. Agric. Food Chem. 2019; 67: 11420-11427. doi: 10.1021/acs.jafc.9b04816

[105] Psaltopoulou T, Kosti RI, Haidopoulos D, Dimopoulos M, Panagiotakos DB. Olive oil intake is inversely related to cancer prevalence: A systematic review and a meta-analysis of 13,800 patients and 23,340 controls in 19 observational studies. Lipids Health Dis. 2011; 10: 127. doi: 10.1186/1476-511X-10-127

[106] Tambalis KD, Panagiotakos DB, Psarra G, Sidossis LS. Exclusive olive oil consumption was favorably associated with metabolic indices and lifestyle factors in schoolchildren. Nutrition, Metabolism and Cardiovascular Diseases. 2019; 30(14): 566-573. doi.org/10.1016/j.numecd.2019.12.004

[107] Saibandith B, Spencer JPE, Rowland IR, Commane DM. Olive Polyphenols and the Metabolic Syndrome. Molecules. 2017; 22: 1082. doi:10.3390/molecules22071082

[108] Finicelli M, Squillaro T, Di Cristo F, Di Salle A, Melone MAB, Galderisi U, Peluso G. Metabolic syndrome, Mediterranean diet, and polyphenols: Evidence and perspectives. J Cell Physiol. 2019;234:5807-5826. DOI: 10.1002/jcp.27506

[109] Storniolo CE, Casillas R, Bulló M, Castañer O, Ros E, Sáez GT, Toledo E, Estruch R, Ruiz-Gutiérrez V, Fitó M, et al. A Mediterranean diet supplemented with extra virgin olive oil or nuts improves endothelial markers involved in blood pressure control in hypertensive women. Eur. J. Nutr. 2017; 56: 89-97. doi: 10.1007/s00394-015-1060-5

[110] Franquesa M, Pujol-Busquets G, García-Fernández E, Rico L, Shamirian-Pulido L, Aguilar-Martínez A, Medina FX, Serra-Majem L, Bach-Faig A. Mediterranean Diet and Cardiodiabesity: A Systematic Review

through Evidence-Based Answers to Key Clinical Questions. *Nutrients*. 2019; 11(3):655. https://doi.org/10.3390/nu11030655

[111] Loizzo MR, Lecce GD, Boselli E, Menichini F, Frega NG. Inhibitory Activity of Phenolic Compounds FromExtraVirgin Olive Oils on the Enzymes Involved in Diabetes, Obesity and Hypertension. J. Food Biochem. 2011; 35: 381-399. doi.org/10.1111/j.1745-4514.2010.00390.x

[112] Storniolo CE, Casillas R, Bulló M, Castañer O, Ros E, Sáez GT, Toledo E, Estruch R, Ruiz-Gutiérrez V, Fitó M, Martínez-González MA, Salas-Salvadó J, Mitjavila MT, Moreno JJ. A Mediterranean diet supplemented with extra virgin olive oil or nuts improves endothelial markers involved in blood pressure control in hypertensive women. Eur. J. Nutr. 2017; 56: 89-97. doi: 10.1007/s00394-015-1060-5

[113] Martin-Pelaez S, Mosele JI, Pizarro N, Farràs M, Rafael de la Torre R, Subirana I, Perez-Cano FJ, Castañer O, Solà R, Fernandez-Castillejo S, Heredia S, Farré M, Motilva MJ, Fitó M. Effect of virgin olive oil and thyme phenolic compounds on blood lipid profile: implications of human gut microbiota. Eur J Nutr. DOI 10.1007/s00394-015-1063-2

[114] Tsigalou C, Tsolou A, Konstantinidis T, Zafiriou E, Efthymios D, Tsirogianni A, Bogdanos D. Interplay between Mediterranean Diet and Gut Microbiota in the Interface of Autoimmunity: An Overview. Preprints (www.preprints.org). 2020. doi:10.20944/preprints202001.0228.v1

[115] Luisi MLE, Lucarini L, Biffi B, Rafanelli E, Pietramellara G, Durante M, Vidali S, Provensi G, Madiai S, Gheri CF, Masini E and Ceccherini MT. Effect of Mediterranean Diet Enriched in High Quality Extra Virgin Olive Oil on Oxidative Stress, Inflammation and Gut Microbiota in Obese and Normal Weight Adult Subjects. Front. Pharmacol. 2019; 10:1366. doi: 10.3389/fphar.2019.01366

[116] Razquin C, Martinez JA, Martinez-Gonzalez MA, Mitjavila MT, Estruch R,Marti A. A 3 years follow-up of a Mediterranean diet rich in virgin olive oil is associated with high plasma antioxidant capacity and reduced body weight gain. European Journal of Clinical Nutrition. 2009; 63: 1387-1393

[117] Parkinson L, Cicerale S. The Health Benefiting Mechanisms of Virgin Olive Oil Phenolic Compounds. Molecules. 2016; 21: 1734. doi:10.3390/molecules21121734

[118] Santangelo C, Vari R, Scazzocchio B, De Sanctis P, Giovannini C, D'Archivio M, Masella R. Anti-inflammatory Activity of Extra Virgin Olive Oil Polyphenols: Which Role in the Prevention and Treatment of Immune-Mediated Inflammatory Diseases? Endocr Metab Immune Disord Drug Targets. 2018; 18(1): 36-50. doi: 10.2174/1871530317666171114114321

[119] M. Aparicio-Soto, M. Sanchez-Hidalgo, M. A. Rosillo, M. L. Castejón and C. Alarcon-de-la-Lastra, Food Funct., 2016, DOI: 10.1039/C6FO01094F

[120] Aparicio-Soto M, Redhu D, Sánchez-Hidalgo M, Fernández-Bolaños JG, Alarcón-de-la-Lastra C, Worm M, Babina M. Olive-Oil-Derived Polyphenols Effectively Attenuate Inflammatory Responses of Human Keratinocytes by Interfering with the NF-kB Pathway. Mol. Nutr. Food Res. 2019; 63(21): e1900019. doi: 10.1002/mnfr.201900019

Chapter 2

Anti-Cancer and Cardiovascular Properties of Phenolic Compounds Present in Virgin Olive Oil

Mohammed El Haouari

Abstract

Cancer and cardiovascular diseases (CVD) are the leading cause of death worldwide. Furthermore, current cardiovascular and cancer therapy is accompanied by various side effects, which considerably reduce the quality of life. Epidemiological studies suggest that the Mediterranean diet has been related to a lower risk of non-communicable diseases such as CVD and cancer. This lower incidence has been partially attributed to the regular intake of virgin olive oil (VOO) which is the main fatty component of the traditional Mediterranean die. In addition to monounsaturated fatty acid, VOO contains various phenolic compounds, which have shown a broad spectrum of pharmacological properties due to their antioxidant activity. This chapter summarizes current knowledge on the effects of the main phenolic compounds isolated from VOO on different cancers and CVD as well as the plausible action mechanisms involved.

Keywords: olive oil, phenolic compounds, cancer, cardiovascular disease, health benefits

1. Introduction

Cardiovascular diseases (CVD) and cancer, which are the main causes of morbidity and death worldwide, occurring as a result of complex factors such as stress, hypertension, hypercholesterolemia, obesity, smoking, inadequate diet and physical inactivity [1]. In addition, current anti-cancer and cardiovascular therapy is based on conventional drugs that have limited effectiveness and adverse side effects such as toxicity of chemotherapeutic drugs on normal cells and reduction of the quality of life. Thus, the need for newer and more effective drugs for the management of cancer and CVD is of great interest. Several epidemiological studies have shown a lower incidence of CVD and certain kinds of cancer in the Mediterranean region due to the Mediterranean diet (MD) which is rich in vegetables, cereals, fruit, fish, and olive oil [2]. In addition, several epidemiological studies suggested that olive oil intake is involved in preventing various cancers as well as CVD mortality and incidence [3–7]. In this sense, a clinical study by Estruch et al. reported that the consumption of 50 mL/day of extra-virgin olive oil (EVOO) reduced the incidence of major cardiovascular events [8]. Besides the traditional benefits on the high level of monounsaturated fatty acid provided by olive oil intake, a broad spectrum of benefits on cardiovascular risk factors and cancer is now emerging associated with olive oil consumption [9]. Thus, the anticancer and cardioprotective properties of olive oil seem to correlate with the antioxidant and anti-inflammatory activity of multiple minor components such as hydroxytyrosol, tyrosol, and their secoiridoid derivatives. Further research have shown that olive oil phenolic compounds exert a possible chemoprotective and anticancer effects in different types of cancers such as breast cancer [10, 11], colon [12], prostate [13] and melanoma [14]. These phenolic compounds exist mainly in extra virgin and VOO and give the oil its health properties. The huge number and variety of phenolic compounds in olive oil might explain the unique health benefits of this culinary oil [15].

2. Virgin olive oil in the Mediterranean diet

The traditional MD contains a considerable proportion of fruits, vegetables, cereals, fish, milk and olive oil. VOO represents the main dietary source of fat in the MD [16]. Olive oil is a flavorsome, tasty and nutritious edible fat obtained directly from pressing ripe olives. Intake of olive oil in the Mediterranean countries is estimated to be 30–50 g/day in Greece, Italy, and Spain. Evidence supports the hypothesis that the health benefits properties of the MD and its ability to reduce the incidence of some degenerative diseases, such as CVD and cancer, may be attributed, at least in part, to VOO [5]. Historically, the high nutritional quality and the health benefits effects of EVOO intake were first attributed to the high concentration of monounsaturated fatty acids [17, 18]. However, greater attention has recently focused on a fraction of minor components (about 2% of oil weight), such as phenolic compounds which have a strong antioxidant activity [19]. These compounds are responsible for EVOO oxidative stability and sensorial properties (such as bitterness and pungency) [20].

3. Phenolic compounds present in virgin olive oil

Olive oil is composed of two fractions, the saponificable and the nonsaponificable fraction. The main constituents of the saponificable fraction are triglycerides (98–99%). The nonsaponificable fraction (1–2%) contains the minor constituents of olive oil such as tocopherols, sterols, chlorophylls, carotenoids, alcohols, waxes, aldehydes, esters, ketones and phenolic compounds [21].

The beneficial effects of the MD are mainly attributed to the antioxidant property of olive oil phenolic compounds. EVOO contains much higher amounts of polyphenols than common olive oil. The phenolic concentration in EVOO ranges from 50 to 800 mg/kg [22]. Phenolic compounds identified in EVOO includes three categories: simple phenols (such as vanillic, gallic, coumaric and caffeic acids, tyrosol and hydroxytyrosol), secoiridoids (such as oleuropein, oleocanthal, and oleacein), and lignans (1-acetoxypinoresinol and pinoresinol) [23]. Secoiridoid derivatives are the most abundant phenols in olive oil. Hydroxytyrosol and tyrosol, which originate from the hydrolysis of oleuropein, are part of the phenolic alcohol group. There is approximately 2 grams of hydroxytyrosol per 100 grams of olive [24]. The concentration of hydroxytyrosol and tyrosol increases as the fruits ripen, in parallel with the hydrolysis of compounds of higher molecular weight, while the total amount of phenolic compounds and α-tocopherol decreases as the fruits ripen [18, 25–27]. Hydroxytyrosol and tyrosol have been the subject of numerous investigations. To date, more than 36 phenolic compounds have been isolated from

EVOO and identified, although they are present at very different concentrations (0.02–600 mg/kg) [28]. Oleuropein is the main polyphenol found in olive oil, both in this form and as the aglycone. It has been shown that the oleuropein content is higher in the first stages of fruit maturation and in green cultivars than in black olives [18]. The phenolic compounds are mainly responsible for the organoleptic characteristics (aroma and flavor) [29, 30] and oxidative stability of the olive oil [31, 32]. The concentration of phenolic compounds in VOO is influenced by a number of factors such as the area of growth, the climate, the variety, the ripeness of the olive, the olive storage and maturation conditions, the production process and the olive tree age [33, 34].

4. Cardiovascular properties of olive oil phenolic compounds

Cardiovascular diseases are a leading cause of death and disability worldwide. They include a group of disorders such as coronary heart disease (CHD), cerebrovascular disease, peripheral arterial disease, pulmonary embolism, rheumatic heart disease, and congenital heart disease and deep vein thrombosis. Several cardiovascular risk factors, such as hypertension, dyslipidemia, diabetes mellitus, obesity and smoking, cause endothelial dysfunction, which lead to the onset of the inflammatory process in atherosclerosis [35]. Remarkably, these pathologies can be largely preventable since unhealthy diet contributes to nearly 80% of risk factors [36, 37]. In this sense, the low rate of cardiovascular mortality found in southern European-Mediterranean countries, in comparison with other westernized countries, despite a high prevalence of coronary heart disease risk factors could be attributed to the olive oil consumption. In addition, numerous randomized clinical trials have shown that consumption of olive oil is associated with beneficial effects on different cardiovascular biomarkers, such as blood lipids, blood pressure, inflammation and thrombosis [38]. Many authors have reported that VOO exerts a preventive effect against CVD [4, 9, 39] by improving many CVD risk factors including blood pressure, glucose metabolism and antithrombotic profile [40]. As evidenced by various scientific studies, the major beneficial properties of VOO have been attributed to the antioxidant of its phenolic compounds. In fact, it has been shown that increased oxidative stress is associated to the pathogenesis of various risk factors of CVD including hypertension, diabetes, platelet hyperactivity and atherosclerosis [41]. Guasch-Ferré et al. [40] found that greater consumption of total olive oil, especially EVOO (rich in phenolic compounds repetition), was associated with reduced CVD: each 10 g/d increase in total olive oil consumption is associated with a 16% reduction in cardiovascular mortality. Low-density lipoprotein (LDL) oxidation plays a critical role in the development of atherosclerosis and coronary heart disease. Various *in vitro* and *in vivo* studies have shown that the polyphenolic compounds of EVOO play an important role in the prevention of atherosclerotic damage through their inhibition of LDL oxidation [42–45].

Another important risk factor for the onset of atherosclerosis is a high blood concentration of cholesterol and particularly LDL cholesterol. The 3-hydroxy 3-methylglutaryl (HMG)-CoA reductase is an enzyme involved in the synthesis of cholesterol. Some studies have focused attention on the effect of the polyphenolic compounds from VOO on the activity of HMG-CoA reductase. It has been demonstrated that the activity of HMG-CoA reductase decreased significantly in the liver microsomes of rats fed with polyphenols extracted from VOO [46]. The inhibition of HMG-CoA reductase by VOO phenolic compounds may thus play an important role in the prevention of CVD. Other research has shown that hydroxytyrosol can prevent CVD by reducing the expression of adhesion molecules on endothelial cells and preventing the oxidation of LDL [47].

Summarizing, the data indicate that olive oil phenolic compounds are associated with a beneficial impact on CVDs. Clinical studies confirm these beneficial effects as shown by the reduction by olive oil polyphenols intake of the inflammatory processes involved in degenerative and chronic diseases such as CVD and cancer.

5. Anti-cancer effects and molecular mechanisms induced by olive oil phenolic compounds

Cancer is a growing health concern worldwide, especially associated with unhealthy lifestyle and physical inactivity. Natural compounds can provide a real benefit as a chemopreventive and/or treatment of this complex disease. EVOO seems to have a protective effect against cancer. Actually, different authors have stated the cancer lowered incidences to olive oil intake, at least in part. The anticancer properties of olive oil polyphenols have been confirmed in several studies. In fact, several epidemiologic studies demonstrated an inverse association between the consumption of olive oil and a reduced risk of different types of cancers such as breast [48–50], prostate [51], lung [52], laryngeal [53], colorectal [54] and ovarian [55] cancer. Numerous studies have investigated the anticancer effect of olive oil phenolic compounds using different experimental models. For instance, the olive oil phenolic (−)-oleocanthal (OLCT), a natural compound present in EVOO, has recently been reported to exert anti-cancer activity in a variety of human cancer types. Recently, our research group investigated the role of OLCT in the development of different cancer hallmarks including proliferation and migration in triple negative MDA-MB-231 cells, as well as in the regulation of intracellular Ca^{2+} homeostasis. Our results indicate that OLCT induced selective anti-proliferative and anti-migrative effect on the triple negative MDA-MB-231 and the luminal MCF7 cell lines, without having any effect on the non-tumoral MCF10A cells. Furthermore, we demonstrated for the first time selective activation of TRPC6-dependent Ca^{2+} influx and TRPC6 downregulation by olive oil-derived OLCT in breast cancer cell lines, which might be responsible for the inhibitory effects of OLCT on cell proliferation and migration [56]. Additionally, several lines of evidence suggest that OLC is active against different types of cancers including hepatocellular carcinoma, multiple myeloma and breast, prostate and pancreatic cancer [10, 57, 58]. The anticancer molecular mechanisms of OLC (comprehensively reviewed in [59]) may involve various cellular signaling pathways such as modulation of the apoptotic pathway and inhibition of the HGF/c-Met and the STAT3 signaling pathways. Mechanisms of anticancer activity of olive oil phenolic compounds are summarized in **Figure 1**.

A recent trial evaluated the *in vitro* anticancer and chemopreventive potential of two EVOO extracts (tyrosol and hydroxytyrosol) and secoiridoid derivatives (oleocanthal and oleacein) on cutaneous non-melanoma skin cancer models. Results demonstrated that phenolic EVOO extracts can block molecular steps that occur after the initial UV radiation exposure and before or during tumor development. In particular, these results indicated that secoiridoid derivatives contribute more than simple phenols to the mechanism of action of EVOO extracts [60]. In a recent review, Farooqi et al. reported that oleuropein acts as an anticancer agent by several cellular mechanisms, such as targeting HER2, epigenetic modifications,

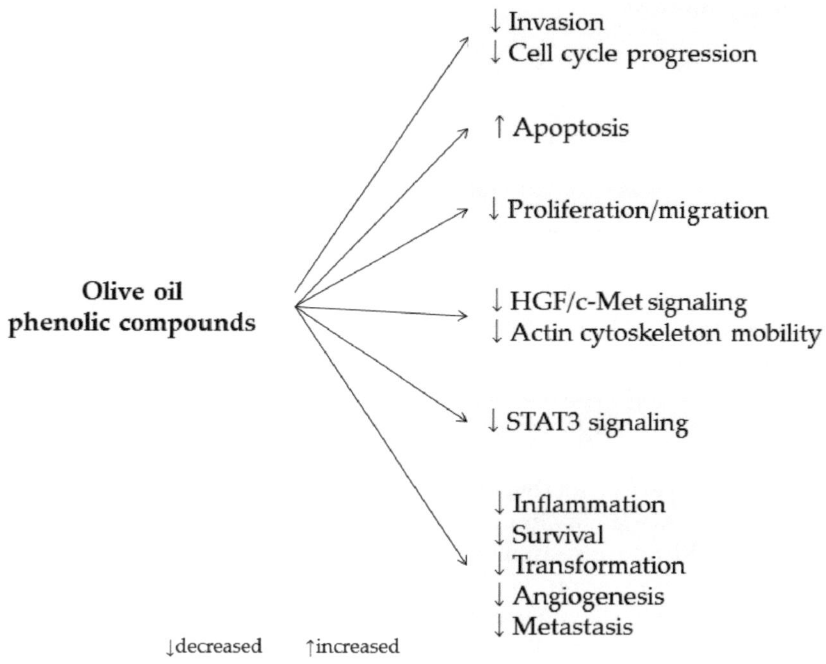

Figure 1.
Anticancer mechanisms of phenolic compounds from olive oil.

interfering with MAPK pathway, modulation of apoptosis and PI3K/AKT signaling axis as well as by reducing ROS production in different cell types [61]. Furthermore, it has been demonstrated that hydroxytyrosol induced apoptosis and cell cycle arrest in cancer cells [47].

6. Conclusion

The protective and preventive benefits of regular consumption of olive oil as part of a healthy diet have been largely documented by various scientific works. As an important component of the MD, olive oil has shown to provide more beneficial health effects to those induced by other vegetable oils.

The literature data showed that normal consumption of EVOO or VOO, which is rich in bioactive compounds, is associated to lower incidence of numerous non-communicable ailments such as CVD and certain types of cancer. The pharmacological properties of olive oil have been attributed not only to its high content of triglycerides but also to the minor fraction of polyphenolic compounds such as tyrosol, hydroxytyrosol, oleuropein and oleocanthal. In fact, these compounds have been shown to exert protective and preventive action against CVD and cancer, particularly through their potent antioxidant and anti-inflammatory activity. Furthermore, olive oil phenolic compounds have been reported to reduce several risk factors of CVD and cancer such as diabetes mellitus, hypertension, dyslipidemia and obesity.

As we have seen, the protective and the preventive effects of olive oil and/or its phenolic compounds against CVD and cancer seem to be pretty encouraging. However, more *in vivo* studies and clinical trials are still necessary to verify the beneficial effect of olive oil on human health.

Acknowledgements

This research received no external funding.

Conflict of interest

The author declares no conflict of interest related to this research.

Abbreviations

CVD	Cardiovascular disease
EVOO	Extra-virgin olive oil
HGF	Hepatocyte growth factor
HMG-CoA	Hydroxy-méthyl-glutaryl-coenzyme A
LDL	Low-density lipoprotein
MD	Mediterranean diet
OLCT	(−)-oleocanthal
STAT3	Signal transducer and activator of transcription 3
VOO	Virgin olive oil

Author details

Mohammed El Haouari[1,2]

1 Department of Biology and Earth Sciences, Laboratory of Pedagogical Engineering and Didactics of Sciences and Mathematics (IPDSM), Regional Center for Education Careers and Training (CRMEF Fès-Meknès), Taza, Morocco

2 Laboratory of Natural Substances, Pharmacology, Environment, Modeling, Health and Quality of Life (SNAMOPEQ), Polydisciplinary Faculty of Taza, Sidi Mohamed Ben Abdellah University of Fez, Taza, Morocco

*Address all correspondence to: elhouarim@yahoo.fr

References

[1] Mack M, Gopal A. Epidemiology, Traditional and Novel Risk Factors in Coronary Artery Disease. Heart failure clinics. 2016;12:1-10. doi:10.1016/j.hfc.2015.08.002

[2] Stefani M, Rigacci S. Beneficial properties of natural phenols: highlight on protection against pathological conditions associated with amyloid aggregation. BioFactors. 2014;40;482-493. doi:10.1002/biof.1171

[3] Buckland G, Travier N, Barricarte A, Ardanaz E, Moreno-Iribas C, Sanchez MJ, Molina-Montes E, Chirlaque MD, Huerta JM, Navarro C. et al. Olive oil intake and CHD in the European Prospective Investigation into Cancer and Nutrition Spanish cohort. The British journal of nutrition. 2012;108:2075-2082. doi:10.1017/S000711451200298X.

[4] Bendinelli B, Masala G, Saieva C, Salvini S, Calonico C, Sacerdote C, Agnoli C, Grioni S, Frasca G, Mattiello A et al. Fruit, vegetables, and olive oil and risk of coronary heart disease in Italian women: the EPICOR Study. The American journal of clinical nutrition. 2011;93:275-283. doi:10.3945/ajcn.110.000521.

[5] Owen RW, Giacosa A, Hull WE, Haubner R, Wurtele G, Spiegelhalder B, Bartsch H. Olive-oil consumption and health: the possible role of antioxidants. The Lancet. Oncology. 2000;1:107-112. doi:10.1016/s1470-2045(00)00015-2.

[6] Psaltopoulou T, Kosti R.I, Haidopoulos D, Dimopoulos M, Panagiotakos D.B. Olive oil intake is inversely related to cancer prevalence: a systematic review and a meta-analysis of 13,800 patients and 23,340 controls in 19 observational studies. Lipids in health and disease. 2011;10:127. doi:10.1186/1476-511X-10-127.

[7] Pelucchi C, Bosetti C, Negri E, Lipworth L, La Vecchia C. Olive oil and cancer risk: an update of epidemiological findings through 2010. Current pharmaceutical design. 2011;17:805-812. doi:10.2174/138161211795428920.

[8] Estruch R, Ros E, Salas-Salvado J, Covas MI, Corella D, Aros F, Gomez-Gracia E, Ruiz-Gutierrez V, Fiol M, Lapetra J. et al. Primary prevention of cardiovascular disease with a Mediterranean diet. The New England journal of medicine. 2013;368:1279-1290. doi:10.1056/NEJMoa1200303.

[9] Covas MI, Konstantinidou V, Fito M. Olive oil and cardiovascular health. Journal of cardiovascular pharmacology. 2009;54:477-482 doi:10.1097/FJC.0b013e3181c5e7fd.

[10] Khanfar MA, Bardaweel SK, Akl MR, El Sayed KA. Olive Oil-derived Oleocanthal as Potent Inhibitor of Mammalian Target of Rapamycin: Biological Evaluation and Molecular Modeling Studies. Phytotherapy research : PTR. 2015;29:1776-1782. doi:10.1002/ptr.5434.

[11] Calahorra J, Martinez-Lara E, De Dios C, Siles E. Hypoxia modulates the antioxidant effect of hydroxytyrosol in MCF-7 breast cancer cells. PloS one. 2018;13: e0203892. doi:10.1371/journal.pone.0203892.

[12] Hashim YZ, Worthington J, Allsopp P, Ternan NG, Brown EM, McCann MJ, Rowland IR, Esposto S, Servili M, Gill CI. Virgin olive oil phenolics extract inhibit invasion of HT115 human colon cancer cells in vitro and in vivo. Food & function. 2014;5: 1513-1519. doi:10.1039/c4fo00090k.

[13] Zubair, H, Bhardwaj A, Ahmad A, Srivastava SK, Khan MA, Patel GK,

Singh S, Singh AP. Hydroxytyrosol Induces Apoptosis and Cell Cycle Arrest and Suppresses Multiple Oncogenic Signaling Pathways in Prostate Cancer Cells. Nutrition and cancer. 2017;69:932-942. doi:10.1080/01635581.2017.1339818.

[14] Song H, Lim DY, Jung JI, Cho HJ, Park, SY, Kwon GT, Kang, YH, Lee KW, Choi MS, Park JHY. Dietary oleuropein inhibits tumor angiogenesis and lymphangiogenesis in the B16F10 melanoma allograft model: a mechanism for the suppression of high-fat diet-induced solid tumor growth and lymph node metastasis. Oncotarget. 2017;8:32027-32042. doi:10.18632/oncotarget.16757.

[15] Bravo L. Polyphenols: chemistry, dietary sources, metabolism, and nutritional significance. Nutrition reviews. 1998;56:317-333. doi:10.1111/j.1753-4887.1998.tb01670.x.

[16] Bach-Faig A, Berry EM, Lairon D, Reguant J, Trichopoulou A, Dernini S, Medina FX., Battino M, Belahsen R, Miranda G. et al. Mediterranean diet pyramid today. Science and cultural updates. Public health nutrition. 2011;14:2274-2284. doi:10.1017/S1368980011002515.

[17] Corona G, Spencer JP, Dessi MA. Extra virgin olive oil phenolics: absorption, metabolism, and biological activities in the GI tract. Toxicology and industrial health. 2009;25:285-293. doi:10.1177/0748233709102951.

[18] Tripoli E, Giammanco M, Tabacchi G, Di Majo D, Giammanco S, La Guardia M. The phenolic compounds of olive oil: structure, biological activity and beneficial effects on human health. Nutrition research reviews. 2005;18:98-112. doi:10.1079/NRR200495.

[19] Vissers MN, Zock PL, Katan MB. Bioavailability and antioxidant effects of olive oil phenols in humans: a review. European journal of clinical nutrition. 2004;58;955-965. doi:10.1038/sj.ejcn.1601917.

[20] Bendini A, Cerretani L, Carrasco-Pancorbo A, Gomez-Caravaca AM, Segura-Carretero A, Fernandez-Gutierrez A, Lercker G. Phenolic molecules in virgin olive oils: a survey of their sensory properties, health effects, antioxidant activity and analytical methods. An overview of the last decade. Molecules. 2007;12:1679-1719. doi:10.3390/12081679.

[21] Aiello A, Guccione GD, Accardi G, Caruso C. What olive oil for healthy ageing? Maturitas. 2015;80:117-118. doi:10.1016/j.maturitas.2014.10.016.

[22] Pedan V, Popp M, Rohn S, Nyfeler M, Bongartz A. Characterization of Phenolic Compounds and Their Contribution to Sensory Properties of Olive Oil. Molecules 2019;24. doi:10.3390/molecules24112041.

[23] Servili M, Montedoro G. Contribution of phenolic compounds to virgin olive oil quality. Eur J Lipid Sci Technol. 2002;104:602-613.

[24] Marsilio V, Campestre C, Lanza B. Phenolic compounds change duringCalifornia-style ripe olive processing. Food Chem. 2001;74:55-60.

[25] Esti M, Cinquanta L, Notte E, La Notte E. Phenolic compounds in different olive varieties. Journal of Agricultural and Food Chemistry. 1998;46:32-35.

[26] Gutierrez F, Jimenez B, Ruiz A, Albi M. Effect of olive ripeness on the oxidative stability of virgin olive oil extracted from the varieties picual and hojiblanca and on the different components involved. Journal of Agricultural and Food Chemistry. 1999;47:121-127.

[27] Brenes M, Garcia A, Garcia P, Rios J, Garrido A. Phenolic compounds in Spanish olive oils. Journal of Agricultural and Food Chemistry. 1999;47:3535-3540.

[28] Cicerale S, Lucas LJ, Keast RS. Antimicrobial, antioxidant and anti-inflammatory phenolic activities in extra virgin olive oil. Current opinion in biotechnology. 2012;23;129-135; doi:10.1016/j.copbio.2011.09.006.

[29] Valli E, Bendini A, Popp M, Bongartz A. Sensory analysis and consumer acceptance of 140 high-quality extra virgin olive oils. Journal of the science of food and agriculture 2014, 94, 2124-2132, doi:10.1002/jsfa.6535.

[30] Procida G, Cichelli A, Lagazio C, Conte LS. Relationships between volatile compounds and sensory characteristics in virgin olive oil by analytical and chemometric approaches. Journal of the science of food and agriculture. 2016;96:311-318, doi:10.1002/jsfa.7096.

[31] Montano A, Hernandez M, Garrido, I, Llerena JL, Espinosa F. Fatty Acid and Phenolic Compound Concentrations in Eight Different Monovarietal Virgin Olive Oils from Extremadura and the Relationship with Oxidative Stability. International journal of molecular sciences. 2016;17: doi:10.3390/ijms17111960.

[32] Del Carlo M, Sacchetti G, Di Mattia C, Compagnone D, Mastrocola D, Liberatore L, Cichelli A. Contribution of the phenolic fraction to the antioxidant activity and oxidative stability of olive oil. Journal of agricultural and food chemistry. 2004;52: 4072-4079. doi:10.1021/jf049806z.

[33] Cicerale S, Conlan XA, Sinclair AJ, Keast RS. Chemistry and health of olive oil phenolics. Critical reviews in food science and nutrition. 2009;49:218-236. doi:10.1080/10408390701856223.

[34] Hachicha Hbaieb R, Kotti F, Gargouri M, Msallem M, Vichi S. Ripening and storage conditions of Chetoui and Arbequina olives: Part I. Effect on olive oils volatiles profile. Food chemistry. 2016;203:548-558. doi:10.1016/j.foodchem.2016.01.089.

[35] Libby P, Ridker PM, Hansson GK. Progress and challenges in translating the biology of atherosclerosis. Nature. 2011;473:317-325. doi:10.1038/nature10146.

[36] Yusuf S, Hawken S, Ounpuu S, Dans T, Avezum A, Lanas F, McQueen M, Budaj A, Pais P, Varigos J. et al. Effect of potentially modifiable risk factors associated with myocardial infarction in 52 countries (the INTERHEART study): case-control study. Lancet. 2004;364:937-952. doi:10.1016/S0140-6736(04)17018-9.

[37] Stampfer MJ, Hu FB, Manson JE, Rimm EB, Willett WC. Primary prevention of coronary heart disease in women through diet and lifestyle. The New England journal of medicine. 2000;343:16-22. doi:10.1056/NEJM200007063430103.

[38] Lopez-Miranda J, Perez-Jimenez F, Ros E, De Caterina R, Badimon L, Covas MI, Escrich E, Ordovas JM, Soriguer F, Abia R. et al. Olive oil and health: summary of the II international conference on olive oil and health consensus report, Jaen and Cordoba (Spain) 2008. Nutrition, metabolism, and cardiovascular diseases : NMCD. 2010;20:284-294. doi:10.1016/j.numecd.2009.12.007.

[39] Romani A, Ieri F, Urciuoli S, Noce A, Marrone G, Nediani C, Bernini R. Health Effects of Phenolic Compounds Found in Extra-Virgin Olive Oil, By-Products, and Leaf of *Olea europaea* L. Nutrients. 2019;11: doi:10.3390/nu11081776.

[40] Guasch-Ferre M, Hu FB, Martinez-Gonzalez MA, Fito M, Bullo M, Estruch R, Ros E, Corella D, Recondo J, Gomez-Gracia E. et al. Olive oil intake and risk of cardiovascular disease and mortality in the PREDIMED Study. BMC medicine. 2014;12:78. doi:10.1186/1741-7015-12-78.

[41] El Haouari M. Platelet Oxidative Stress and its Relationship with Cardiovascular Diseases in Type 2 Diabetes Mellitus Patients. Current medicinal chemistry 2019, 26, 4145-4165. doi:10.2174/0929867324666171005114456.

[42] Fito M, Covas MI, Lamuela-Raventos RM, Vila J, Torrents L, de la Torre C, Marrugat J. Protective effect of olive oil and its phenolic compounds against low density lipoprotein oxidation. Lipids. 2000;35:633-638. doi:10.1007/s11745-000-0567-1.

[43] Visioli F, Bellomo G, Montedoro G, Galli C. Low density lipoprotein oxidation is inhibited in vitro by olive oil constituents. Atherosclerosis. 1995;117:25-32. doi:10.1016/0021-9150(95)05546-9.

[44] Rice-Evans CA, Miller NJ, Paganga G. Structure-antioxidant activity relationships of flavonoids and phenolic acids. Free radical biology & medicine. 1996;20:933-956. doi:10.1016/0891-5849(95)02227-9.

[45] Masella R, Cantafora A, Modesti D, Cardilli A, Gennaro L, Bocca A, Coni E. Antioxidant activity of 3,4-DHPEA-EA and protocatechuic acid: a comparative assessment with other olive oil biophenols. Redox report : communications in free radical research. 1999;4:113-121. doi:10.1179/135100099101534792.

[46] Benkhalti FJP, Paz E, Perez-Jimenez F, El Modafar C, El Boustani E. Effects of feeding virgin olive oil or their polyphenols on lipid of rat liver. Nutrition Research; 2002;22:1067-1075.

[47] Rafehi H, Ververis K, Karagiannis TC. Mechanisms of action of phenolic compounds in olive. Journal of dietary supplements. 2012;9:96-109. doi:10.3109/19390211.2012.682644.

[48] Martin-Moreno JM, Willett WC, Gorgojo L, Banegas JR, Rodriguez-Artalejo F, Fernandez-Rodriguez JC, Maisonneuve P, Boyle P. Dietary fat, olive oil intake and breast cancer risk. International journal of cancer. 1994;58:774-780. doi:10.1002/ijc.2910580604.

[49] Trichopoulou A, Katsouyanni K, Stuver S, Tzala L, Gnardellis C, Rimm E, Trichopoulos D. Consumption of olive oil and specific food groups in relation to breast cancer risk in Greece. Journal of the National Cancer Institute; 1995;87:110-116. doi:10.1093/jnci/87.2.110.

[50] la Vecchia C, Negri E, Franceschi S, Decarli A, Giacosa A, Lipworth L. Olive oil, other dietary fats, and the risk of breast cancer (Italy). Cancer causes & control : CCC. 1995;6:545-550. doi:10.1007/BF00054164.

[51] Hodge AM, English DR, McCredie MR, Severi G, Boyle P, Hopper JL, Giles GG. Foods, nutrients and prostate cancer. Cancer causes & control:CCC.2004;15:11-20.doi:10.1023/B:CACO.0000016568.25127.10.

[52] Fortes C, Forastiere F, Farchi S, Mallone S, Trequattrinni T, Anatra F, Schmid G, Perucci CA. The protective effect of the Mediterranean diet on lung cancer. Nutrition and cancer. 2003;46:30-37. doi:10.1207/S15327914NC4601_04.

[53] Bosetti C, La Vecchia C, Talamini R, Negri E, Levi F, Dal Maso L, Franceschi S. Food groups and laryngeal cancer risk: a case-control study from Italy and Switzerland. International

journal of cancer. 2002;100:355-360. doi:10.1002/ijc.10485.

[54] Stoneham M, Goldacre M, Seagroatt V, Gill L. Olive oil, diet and colorectal cancer: an ecological study and a hypothesis. Journal of epidemiology and community health. 2000;54:756-760. doi:10.1136/jech.54.10.756.

[55] Bosetti C, Negri E, Franceschi S, Talamini R, Montella M, Conti E, Lagiou P, Parazzini F, La Vecchia C. Olive oil, seed oils and other added fats in relation to ovarian cancer (Italy). Cancer causes & control : CCC. 2002;13:465-470. doi:10.1023/a:1015760004130.

[56] Diez-Bello R, Jardin I, Lopez JJ, El Haouari M, Ortega-Vidal J, Altarejos J, Salido GM, Salido S, Rosado JA. (−) Oleocanthal inhibits proliferation and migration by modulating Ca(2+) entry through TRPC6 in breast cancer cells. Biochimica et biophysica acta. Molecular cell research. 2019;1866:474-485. doi:10.1016/j.bbamcr.2018.10.010.

[57] Akl MR, Ayoub NM, Mohyeldin MM, Busnena BA, Foudah AI, Liu, YY, Sayed KA. Olive phenolics as c-Met inhibitors: (−)-Oleocanthal attenuates cell proliferation, invasiveness, and tumor growth in breast cancer models. PloS one. 2014;9:e97622. doi:10.1371/journal.pone.0097622.

[58] LeGendre O, Breslin PA, Foster DA. (−)-Oleocanthal rapidly and selectively induces cancer cell death via lysosomal membrane permeabilization. Molecular & cellular oncology. 2015;2:e1006077. doi:10.1080/23723556.2015.1006077.

[59] El Haouari M, Quintero JE, Rosado JA. Anticancer molecular mechanisms of oleocanthal. Phytotherapy research : PTR. 2020;34:2820-2834. doi:10.1002/ptr.6722.

[60] Polini B, Digiacomo M, Carpi S, Bertini S, Gado F, Saccomanni G, Macchia M, Nieri P, Manera C, Fogli S. Oleocanthal and oleacein contribute to the in vitro therapeutic potential of extra virgin oil-derived extracts in non-melanoma skin cancer. Toxicology in vitro : an international journal published in association with BIBRA. 2018, 52, 243-250, doi:10.1016/j.tiv.2018.06.021.

[61] Ahmad Farooqi A, Fayyaz S, Silva AS, Sureda A, Nabavi SF, Mocan A, Nabavi, SM, Bishayee A. Oleuropein and Cancer Chemoprevention: The Link is Hot. Molecules. 2017;22: doi:10.3390/molecules22050705.

Chapter 3

Prospective Adaptation of the Mediterranean Crop Olive in India

Thangamani Dhandapani, K.B. Sridhar and S. Vimala Devi

Abstract

The market for the Mediterranean crop Olive is ever increasing in the Indian Sub-continent. Apart from import, exploring the local possibilities of cultivating Olive in this country is being explored. Adaption and acclimatization of the new crop is always a challenge. Though the country has large areas with similar agronomic regions as that of the native Mediterranean regions, ecological adjustment of the crop to the micro-climate of the new area needs modification of cultural practices. The success of olive cultivation, the challenges encountered, the prospective of making this cultivation a sustainable one by innovative alternate usage is explored in this chapter.

Keywords: olive cultivation, Indian climate

1. Introduction

Olives are one of the world class premium oil producing groups of evergreen trees and shrubs, largely found in warm temperate and tropical regions of the world. It is a unique fruit crop which can withstand high temperatures and drought, but requires chilling temperature for fruiting [1, 2]. Olives can grow in wide range of soils like heavy, light, clayey or sandy soils. It belongs family Oleaceae [2, 3]. All the cultivated varieties of olive are diploid (2n = 46). It is a crucial oil yielding crops of the Mediterranean countries and in sector with similar climate alike Cuba, California, New Zealand and South-East Australia. Its nativity in apparently the eastern region of the Mediterranean Sea, deriving out of position it has circulated about the basin [4, 5]. Olives are grown mainly for two purposes: fruits for pickling and oil extraction. Apart from this, the fine yellow or light greenish brown hard wood is used for furniture making and the leaves have medicinal value. Earlier olives are confined largely to high end consumers, but now it is gaining popularity world over because of its numerous health benefits.

India has observed olives in consideration of the Buddhist periods as per the mention in the Tripitaka include numerous resources of spoiling jaitavans (Olives) by monks afterwards purchasing lands [6]. In 1885, the beginning olive plantation practice was started towards Kashmir, in an Indo-Italian merger. There is also a mention of other experiments included the Indo-Spanish venture for Himachal Pradesh olive plantations. Neither was successful to cultivate large scale olive production.

2. Indian scenario for olives

It is important to note that olive oil is not as costly as recognized in consideration of it is used in 1/3rd the volume of other oils. Also including expands international disclosure, Indians awareness of the assistance of olive oil as edible oil has made consumption of olives and olive oil as a fashion among the middle and the wealthy classes. The During 1970's, olive oil got great boost due to the health benefits and today olive is an important international trade commodity. Between 2009 and 2010, consumption of olive oil in India grew by 52%. Since 2006, it has shown a cumulative growth rate of 30%. The olive has a high content of monounsaturated fat (mainly oleic acid) and polyphenols which are beneficial for health. It is a prosperous source of polyunsaturated fatty acid (PUFA) and is exactly free come out of cholesterol. The consumption of olive oil has been chemically advocated to people suffering from hypertension and coronary disease. The fruit has bitter component (oleuropen), a low sugar content (2.6–6%) compare with other drupes (12% or more) and high oil content (12–30%) depending on time of the year and variety. Moreover, it is also rich in minerals like iron, calcium and phosphorus.

The imports of olive oil are ever growing and India meets 100 percent demand of olive oil through introduction, mainly from Europe. According to Indian olive oil association, the demand of olive oil is growing 20 percent annually. Olive oil is finally creating a niche in India's edible oil market. Due to campaigns like "Olive it Up" and many international companies investing money to develop their brands, the market is widening and Indians are becoming more educated for the quality and use of oils. The 60 percent of the national market is controlled by three companies in India. And 90% of the import is accounted from Spain and Italy. This suggests the potential for cultivation and production of olive oil in the country. Acknowledging the import value of Olive oil, in 2006, the first attempt to cultivate the olives in India was initiated in one district of Rajasthan, but after 15 years, it has spread to five more states. The varieties grown in India are mainly the ones introduced from Egypt, Italy and USA.

3. Olive Research and Development in India

The olive cultivation in India is still in its infancy stage apart from Rajasthan, and restricted to a few pockets in the states of Jammu and Kashmir, Himachal Pradesh and Uttarakhand. Olive is a tree endures oil-seed crops appropriate toward subtropical climatic surroundings, is mostly mature for its oval shape fruit and that is commonly used for extricate non-drying edible oil and along with eaten basic in soups, salads, pickles etc. The National Bureau of Plant Genetic Resources (NBPGR), New Delhi has popularized frequent enhanced olive variation beginning at other region that is healthy well beneath Indian environment. India is importing around 68% of cooking oil each year. There is need to abate the import of cooking oil. Olive oil utilization has risen five times arising out of recent ten years. India has an ample capability for olive planting which needs to be explored. The Regions olive oil imports are developing starting of around 14,000 MT in 2013, imports are estimated to develop up to 42,000 MT by 2025. Olives are high in oleic acid, which can help prevent heart diseases. The major deterrent to greater consumption is its price. One liter of extra virgin can cost between Rs 800 and Rs 1000. Presently, Rajasthan is the dominant state for Olive cultivation in India. Subsequent to, 2013 towards 2016, the state has produced in bulk 11574.09 kilograms of olive oil. At the beginning, the state had initiated the olive cultivation on a total range of 182 hectares on government plantation. The olive cultivation has besides attained up to 425.18 ha on farmer's fields. Seven olive plantation areas are enhancing in different regions of the state. The state had got seven distinct olive category namely Barnea (Origin Israel; purpose: Oil), Arbequina (Origin: Spain; Purpose: Oil), Cortina (Origin Italy; Purpose: Oil), Picholine (Origin France; Purpose: Dual), Picual (Origin Spain; Purpose: Oil), Koroneiki (Origin Greece; Purpose: Oil) &Frantoio (Origin Tuscany, Italy; Purpose: Oil) were imported from Israel. The government is encouraging farmers to plant olives and providing subsidies.

4. The first olive experiments which took off from Rajasthan

The India's tryst along with olive cultivation was established during former Rajasthan Chief Minister Vasundhara Raje visited Israel in 2006. She was wonder struck by the olive trees in Negev desert, an arid landscape much like Rajasthan. She took initiative to green the Indian dessert state Rajasthan and immediately, the Mediterranean trees were imported and planted over 180 hectares of state government land with the help of Israeli technology. Unlike in Mediterranean countries such as Spain and Italy where the trees are old and the farming traditional, the Israeli technique is based on intensive plantation. 1, 12, 339 Saplings of 7 olives varieties *viz* Barnea, Arbequina, Cortina, Koronoiki, Picual, Frontio and Picholine were imported from Israel in 2007. Their rooted cuttings were then hardened at hi-tech nursery in Rajasthan. The plants were irrigated by the latest method i.e., "drip-ferti" irrigation techniques, the method of direct injection, water and nutrients added simultaneously. By 2014–2015, the plants started fruiting in about 13 of the 33 districts of the state. 1300 tones of fruits were harvested in the first season.

After the success of the pilot project, and study of the technical feasibility and economic viability, commercial cultivation of olives started in Rajasthan. Government of Rajasthan determined to encourage the Olive cultivation under public-private partnership in the state and Rajasthan Olive Cultivation Limited (ROCL). For thus commercialism challenge Rajasthan government had united organization along Pune placed inundation tools Finolex Plasson Industries and Indolive an Israel firm with knowledge in olive cultivation in dessert countries. This includes setting up an oil extraction unit, sale and supply of saplings and equipment, and buying back the fruit. The company also set up an olive press in different districts to extract oil. The oil produced from the olive fruits grown at Rajasthan, is as good as international standards. Though there were some bottlenecks of low production, the next phase was initiated giving importance to the positive results (**Figure 1**).

Around 200 hectares under the National Mission on Oilseeds and Oil Palm (NMOOP) and 5000 hectares under the Rashtriya Krishi Vikas Yojana (RKVY) were brought under the olive cultivation. Farmers are being renewed to cultivate olive trees by offering these free plants and mechanical support.

Seven Olive farms had been developed on state owned land in various agro climate zones with a total area of 182-hectare under intensive farming system. Olive Plantation are arrange beneath almost novel and promote methods based on most recent experience – how and action. The methodological know – how is being attained from different authorities which regularly visit, inspect and monitor the progress of farms. Watering is being complete by computer supervised drop watering method. Provinces are well implemented along composed fertigation and climate depot. Different plantation methods likewise – super intensive plantations, fertigation, plant protection, agronomical practices, leaves diagnosis were undertaken.

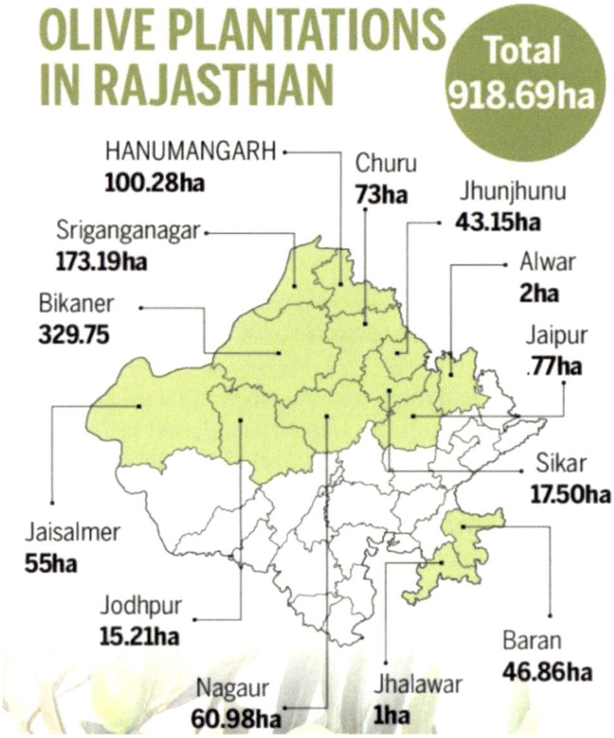

Figure 1.
Olive plantations in Rajasthan, source; ROCL.

5. The olive cultivation at Kashmir, Himachal Pradesh and other states

With the success of Olive cultivation in Rajasthan, Italian companies explored the possibilities at Ramban in Jammu and Kashmir, Kullu in Himachal Pradesh, Punjab, Haryana and Odisha. Some of these attempts failed, as a result, there was a dire need for importing suitable varieties from different olive producing countries and testing for its establishment and large scale cultivation in suitable agro-climatic zones of the country.

All over, for the olive tree, two Olive worlds Germplasm Banks (OWGB) are shortly initiated, in Co'rdoba (Spain) in 1970 and Marrakech (Morocco) in 2003. The olive germplasm bank of Spain has over 350 cultivars collected from different countries. The olives include about 1250 cultivars from over 50 countries. A large number of the olive collection exits in the gene bank of Albania, Algeria, Argentina, Australia, Azerbaijan, Brazil, China, Cyprus, Egypt, France, Greece, India, Iran, Israel, Italy, Japan, Jordan, Montenegro, Morocco, Nepal, Portugal, Slovenia, South Africa, Spain, Tunisia, Turkey and USA [2]. A core collection on olive genetic resources has been developed from 561 accessions from 14 Mediterranean countries [7]. Existence of a good amount of diversity in world olive collection has been reported by several authors [4].

In an effort to import suitable olive varieties for research purposes in the country from various sources, a total of 108 olive varieties were introduced into India by Indian Council for agricultural research- National Bureau of Plant Genetic Resources, New Delhi from USA and Egypt during 2000–2011. The material was provided to the Central Institute of Temperate Horticulture (CITH), Srinagar, Kashmir one of the National Active Germplasm Sites (NAGS) of the ICAR-NBPGR. A low success rate was observed in its establishment and only nine varieties viz.,

Coratina, Frontoio from Egypt and Bouteillon, Dole 0090, Frontoio, Grossa Di Spagna, Leccino, Mission Leiva and Piconia introduced from USA and Egypt survived. Fruiting was observed in three varieties viz., Coratina, Frontoio and Leccino [8].

The Department of Agriculture and Cooperation (DAC), Ministry of Agriculture, India gave permission for bulk import of seed/planting material for cultivation purposes. Therefore, several olive varieties were also introduced into India, by various government and private agencies for cultivation purposes. These varieties were reported to be maintained by the State Horticultural agencies of Himachal Pradesh, Uttarakhand and Jammu and Kashmir. An Indian corporation, in a concerted deal along a Spanish corporation, has declared a project over 300 hectares in Himachal. The Kashmir State is measuring with the 60 varieties attained from the U.S., Egypt and Italy.

Olive is not a native of this country, there is certain dependence for diverse germplasm from major olive producing countries. Olives thrive well in Mediterranean like climate, therefore a certain location in Jammu and Kashmir, Himachal Pradesh, Uttarakhand in India are suited for growing olives in India. A good amount of diversity is present in the olive varieties from Spain, Italy Greece and other countries and there is a potential for growing those varieties in India under suitable agro-climatic region. The systematic germplasm screening and evaluation of olives in India will have significant impact on Indian olive cultivation. In other areas, use of optimum agricultural practices such as drip irrigation is also being attempted for the large scale cultivation. The diverse agro-climatic conditions in India, including the poor and marginal soil can be efficiently utilized for olive production using modern agronomic practices.

6. Cultivation in new areas

With the success of ROCL in Rajasthan, Olives have now caught the fancy of other states. ROCL has distributed4.5 lakhs saplings to Punjab, Himachal Pradesh, Jammu and Kashmir, and Uttar Pradesh and North-Eastern states. It is also helping states such as Haryana, Punjab, and Jammu and Kashmir to set up pilot olive plantations. As per the Indian Olive Association, a federation of olive importers and traders estimates a 40% growth in olive consumption annually. Gujarat, more, is experimentation along 84,000 plants at its cultivation educational institution. Those states need to takings benefits of the commotion in appeal for table olives and olive oil. The Nashik Valley in Maharashtra and the Nandi Hills near Bangalore has similar climate as that of Mediterranean region, hence these pockets also is being explored for olive cultivation.

Under a $300 million project called 'Promotion of Olive Cultivation for Economic Development and Poverty Alleviation,' Italy would help Punjab, Khyber-Pakhtunkhwa, FATA and Baluchistan to grow olive plants.

With the Government of Rajasthan declaring Olive as" Plantation Crop", the Government of India allowed 100% foreign direct investment (FDI) through automotive route. Demonstration of 4D Aqua technology by the ROCL allowed the crop to grow under high salinity. Under these developments, the Government of India has included olive as a crop under National Mission on Oil Seeds and Oil Palm, with these many other states like Arunachal Pradesh, Karnataka, Andhra Pradesh are also coming up to study the feasibility of Olive cultivation in India.

Many companies like Oliva International, is identifying possible land area for olive plantation across India (Rajasthan, Ooty, Himachal Pradesh, Kashmir, Punjab) and are actively exploring new possible areas. It is also providing Indian hardened olive plants with subsidy, design for olive grove and complete end to end value chain consultation.

7. Package and practices for olive cultivation in India

Major Olive producing states: Rajasthan, Uttar Pradesh, Himachal Pradesh and Jammu and Kashmir.

Local Names: Jaitun (Hindi), Aliv/Olipe (Kannada).

Climate: Olive requires warm to subtropical climate. This can be cultivated at an latitude of 1500 m above sea level. The optimum temperature range is 15 to 30 degrees. It requires 100 hours of chilling environment. Mild to cool winters with a chilling period of about 2 months, with average temperatures varying between 1.5–10°C required for flower bud differentiation. It requires long, hot and dry summers to properly maturate the fruit. A few olive category, Similar those grown in Egypt, Tunisia or Israel, blossom and fruit with certain slight winter chilling, whilst other categories expect more chilling for a common flower contrast. Areas receiving a mean annual rainfall of 400 to 700 mm are most suitable for olive growing.

Soil: Olives can be cultivated in wide range of soils. Requires well drained loamy or clay loam soils for best growth. The soil pH range 6–7.5 is ideal for its growth.

Propagation: Olive propagation is made today mainly through semi-hard wood cuttings (60% of the total) and by grafting the desired cv. on olive seedlings (40%). Nowadays micro propagation is also used in olive.

Seeds: very common, however the resulting propagaule is of not true to type.

Grafting/Budding: The seedlings can, nonetheless, be grafted or chip-budded along material coming out of favorable cultivars.

Suckers: Another method of propagation is transplanting suckers that grow at the base of mature trees.

Cuttings: Shoot tips.

Irrigation: Watering is supported specifically in table olive variation where large fruit size is sought. It is further essential in intensive cultivation along massively planted trees for maximum blossoming. Irrigation also enhances the effectiveness of fertilization and pruning. Certainly, it may minimize the incident of substitute bearing. Olive trees are very sensitive to over irrigation and will not perform well in waterlogged soils.

Tree Spacing: the tree spacing ranges from 5 m x 5 m to 6 m x 6 m and 7 m x 7 m. In areas where the climate is especially suitable, more areas are needed because tree development is higher than common.

Choosing the varieties: For early bearing- 'Koroneiki', 'Arbequina', 'Maurino', 'Picual', 'Manzanilla' and for oil quality: 'Frantoio' 'Arbequina', 'Moraiolo', 'Picual' for cold resistance: 'Nostrale di Rigali', 'Leccino', 'Orbetana', 'Dolce Agogia', for Lime tolerance: 'Picudo', 'Cobrançosa', 'Galego', 'Lechín de Sevilla', 'Lechín de Granada', 'Hojiblanca', for salinity tolerance: 'Picual', 'Arbequina', 'Lechín de Sevilla', 'Canivano', 'Nevadillo' .

8. Flowerings and pollination

Flowerings and pollination are especially crucial phase in the olive cultivation. Affluence about flower is the essential of a welfare yield. The appearance of flower in June based on bud appearance that starts on developing shoot tips in April–May of the year back. Flower differentiations appear later. This crucial movement is complicated and introduce along flower bud selection, i.e. the formation of the physiologic action known as nutrient and hormone capability as the bud tips to coming out of the inflorescence axis and flowers. Flower configuration along with organ improvement occurs from March until May–June, though blossom appears. The fruit is produce through fertilization of the egg cell in the flower pistil.

9. Floral and fruit biology

Flower bud inflorescence is endured in the axil of each leaf. Commonly, the bud is produced on the present season's arises and commit conspicuous develop in the coming season. The 300 of flowers each branch. All inflorescence consist of 15 to 30 flower buds, that are short, round and white yellow in color. The flowers are white or whitish; the calyx is small along with four - toothed; the corolla is minute-tubed along four valvate petals; the stamens are two, all bearing 10.000 to 15.000 small also airy pollen grains. The ovary is two- loculed, carry a small style also a capitate stigma. The pistil endures two carpels, all fruit consist of two ovules but one is fertilized also thus formed one-seeded drupe.

10. Fertilization

The dependents total of fertilizers are suggested 1000–1500 kg/ha 0–20-0 and 500–800 kg/ha 0–0-50.These quantity of fertilizers are enough to make up for the need in phosphate also potash as the afterwards 5–8 years. In the later year, next the birth of the new vegetation, 3–4 fertilizations alongside ammonium nitrate (20–30 g/tree every time) are essential followed through watering. The constant is practiced in consecutive years as far as the trees enter the producing point, developing constantly the amount of fertilizer.

11. Intercropping

Olive can be intercropped with wheat and other cereal crops, vegetables such as watermelons, tomatoes and potatoes etc.

12. Prunning and type of cuts

Olive trees should be pruned in preference in wintertime, amidst collecting and sprouting. Pruning should be afterwards in field where wintertime temperature is bottom in that it has a unfavorable event on cool defiance and less temperatures stop damage from curative hurriedly. The cuts made to the branches and shoots should not be too deep to avert notching branches bottom. To aid curative, they should not drop stub.

13. Training

In the meanwhile training, the main purpose of an action is to attain the actual shape as instantly as achievable in regulation to prompt consequent Olive productions. During initiative, accelerated improvement, container developed nursery plants should be used that have mature to acceptable height and have bit side way axis. The soil should implement the developing plants with the greatest attain able environment to mature when they are planted out and consequently. All along this point pruning should be maintained to a minimal to augment developments.

14. Pests

The important pests of crops are the olive fruit fly (Bactroceraoleae), the olive-kernel borer or olive moth (*Prays oleae*) and the black scale (Saissetiaoleae). Even though *B. oleae* is mentioned the common deliberate insect, all three are broadly founded in the Mediterranean countries and appears on olives at population massiveness initiating great economic losses.

15. Diseases

The greatest decisive olive tree diseases are verticillium wilt, olive knot, leaf spot and fruit mummification.

16. Fruit harvest

The fruits can be harvested manually and mechanically. Sometimes fruit may drop naturally.

17. Uses

Ripe olive fruits are pressed for rich oil. The fruit has approximately 20% oil and contains certainly low amount of cholesterol. Olive consists of 80% unsaturated fatty acid in contrast 20% saturated ones. Thus consists of oleic acid in great percentage and that is greatly important as our body. The oil is used as cooking, salad dressing, food preparation, massage, and for the manufacture of cosmetics, Pharmaceuticals, toilet preparation, etc. Matured fruit are also eaten afterwards being treated and dehydrated in vinegar or salt mixture substance. The juicy residue from leaves, bark and fruits contains medicinal components. Its wood is used as carvings, to endow houses, to make vessels and equip for the kitchen as well as for the field benefit and is good firewood. Leaves are food for animals as fodder; oil cake is fed to livestock or is applying for manure. The olive extracts soon afterwards extraction of oil is perfect for manures.

18. Promotion of olive based agroforestry systems

In Greece olive tree AF systems have been considered in intercropping with wheat, barley, maize, and chickpea [9, 10]. In Italy, the imitation of alfalfa (*Medicago sativa* L.) or *Asparagus acutifolius* L. in olive rows provides profitable market niche [11–13]. Durum wheat and chickpeas are common Mediterranean plants, part of the Mediterranean diet and grown under the same environmental conditions as olive trees, they represent interesting candidates to be the associated crops. In Morocco, olive- (*Oleae uropaea* L.) agroforestry is widespread [14] and applied to conventional agriculture condition and proven its resilience over millennia. Several species of trees (e.g., fig, carob, quince), cereals (e.g., wheat, barley), grain legumes (e.g., faba bean, chickpea) grow with/under olive trees [15–17] evaluated two annual crops (durum wheat and faba bean) in olive agroforestry in northern Morocco. Agroforestry improved individual grain weight by 39% for wheat and 17% for faba bean, and enhanced the protein content of wheat grains and straw by 4% and 9%. Olive agroforestry systems can have great LERs and produce high-quality grains, even beneath more arid circumstances than last appraisal in Europe [18].

19. Value addition

To make the olive cultivation remunerative in Indian condition and to fetch good economic returns, new uses were need to explore for the fruitless olive trees. ROCL experimented with olive leaves and has come up with 'Olive tea', which has antioxidant properties and is creating awareness for its health benefits. The olive tea from Rajasthan has found market in UK with a price tag of Rs. 10,000 per kg, which shows the returns are high.

ROCL has also ventured into yet another herbal concoction, i.e., olive wine. Research is being conducted at Vidhyasagar University, West Bengal, Tripura University and Jiwaji University in Indore.

Another alternative, ROCL ventures were initiated was in agro-tourism. Olive tree landscaping is the latest fixation among the affluent. In the vast stretch of sandy terrain of the desert land, Rajasthan, the green stretches of olive plantations are welcomed as resort, based on the theme of olive agri-tourism.

Also another attraction with wood from the tree delicately carved into statuettes is making way in agro-tourism. Most of the demands come from farm houses, hotels, residential societies and high-end bungalows. That's how India, especially Rajasthan, has become an unlikely innovation hub for olives. Today about 150 farmers are growing it on over 900 hectares. Three industrial have come up to formulate the tea.

20. Incentives for olive cultivation

Farmers reach 75 per cent appropriation on plants, 90 per cent appropriation on drop watering also Rs 3000 per hectare as fertilizers along with chemicals. Additionally, we get mechanical assistance of one experienced per 50 ha. Presently, Rajasthan farmers reach one olive plant as Rs 115, although farmers starting with other states buy it at the cost of Rs 150. The government of Rajasthan has declared Olive has plantation crop. Olive cultivation is expected to fetch about 5 times the profit that the farmers of Rajasthan currently fetch from wheat on a hectare land (**Figures 2** and **3**).

Figure 2.
Olive boundary plantation established at farmers field, Deligaon, Jhansi, India.

Figure 3.
Farmer Sahab ram with olives, photo courtesy ROCL.

21. Quality planting material production

An attempt was made to multiply Olives through clonal techniques. The shoot tips collected from one and half year old plantation was used as planting material. The cuttings were quick dipped in IBA 6000 ppm solution and were placed in growing media (vermiculite). The cuttings started rooting after 25 days of planting. Overall cuttings showed 50 percent rooting (**Figure 4**).

Figure 4.
Quality planting material production.

22. Olive oil and market in India

Though*Oleaeuropaea* L. (Olive Tree), grows enormously in Mediterranean countries, the output of the tree is surfing whole world as Olive oil, its main derived product, has experienced an increase in its majority due to its organoleptic attribute and its united useful health effects. Olive oil is considered as the premium product in the oil category whose production is the least in India, since it's not native to India and we are not cultivating this species, so as it is entirely imported from countries like Spain & Italy. In Indian continent, Rajasthan is the only place for the production of olive oil. Olive plants need chilling to flower and fruit the temperature should be below five degrees at night time and below16 to 20 degrees in the morning period. The Indian olive oil market is expected to grow at a CAGR of 9.12% in terms of Value and at a CAGR of 1.05% in terms of volume in coming years.

The India olive oil market size was priced at $58.6 million in 2017, and is projected to reach $127.5 million by 2025, growing at a CAGR of 9.9% from 2018 to 2025. When we analyses the characteristic of the crop oil, it is a viscous broth which is dislodged coming out of the fruit of the olive tree by pressing all olives. Olive oils have less smoke point of 240°Calsocan be consuming natural. This oil utilization is generally investigated healthful for it is united alongside a less difficult of disease of heart and convinced dealings along with colorectal also breast cancer. This oil is also a great origin of monounsaturated fatty acid also antioxidants like polyphenols, vitamins E & K, chlorophyll, also carotenoids. The olive oil corporation in India has endorsed appreciable development in the current years due to acceleration in health-attentive customers. The olive oil market has developed into great competing and expenditure conscious owed to the high capability and constant development of the olive oil marketing. The aspect such as expanding want from end-user companies such as food, personal care, pharmaceuticals, etc. and rising attention about health aspects is expected to drive the India olive oil market growth.

In accord with The Associated Chambers of Commerce and Industry of India (ASSOCHAM), the market size of India's beauty, cosmetic, and grooming production was $6.5 billion in 2016 also is predicated to reach $20 billion by 2025. It's grows in utilization of cosmetic also beauty production are likely to contribute lucrative convenience as the improvement to forthcoming of the India market.

23. Raj olive India's indigenous oil brand

The oil contented in the fruit collected ranged from 9 to 14 per cent. The oil content of the olive in other countries varies between 12 and 16 per cent. The state will be launching its olive oil under the brand 'Raj Olive Oil'. One and other extra virgin and virgin oils are built by automated technique and are not chemically considered. The acceptable common of free fatty acids is greater in a virgin olive oil in comparison to the additional virgin variation. The country's olive oil imports are developing at a quick clip - from around 14,000 MT in 2013, implication are projected to shoot up to 42,000 MT by 2025 as progressively Indians favor olive oil for its healthy aspects. It's plenty as oleic acid amount, which can help prevent heart diseases. According to ROCL, India currently imports 14,500 metric tons of olive oil from countries like Spain, Italy and Greece, and the bottled varieties of extra virgin and pomace olive oil charge amount amid of Rs 700 and Rs 1000 a liter. Although the price of olive oil build upon international charges, we can encourage that indigenously-produced oil will be low cost than those transported.

24. Deterrents to greater oil consumption

The studies have been carried out in the west of the Mediterranean, the following oils such as Arbequina, Carrasquena, Cornicabra, Del Pomet, Hojiblanca, Llumero, Picudo, Seniero, Serrana and some Villalonga contains more than 80% of oleic acid. Lipids are important foods provide nutrition, among the total energy 30% has been intake from the lipids, intake of lipid percent has been dependable with age, weight, and possible sensitivities to pathologies of the consumer, hence among the total lipid intake 12 and 20 per cent has to be monounsaturated fats. Therefore, the active component characterization of monovarietal oils is essential at the moment, there is a clear need to replace the current industrial pastries made with different types of oils and fats, with EVOO product and the daily utilization of amidst 5 and 10 fruits or vegetables. Olive oil utilization has seen develops in the last couple of years as customers are becoming attain of the useful of olive oil management and its aftermath on health. There are three main types of olive oil Extra-virgin, Refined and pomace.

The pomace olive in 2017 assumed as the best contribution in expressions of profit in the Indian olive oil market and is expected to develop at a constant growth rate amid the predicted timings. These are applying to the pomace olive oil being a more comfortably costing and gladly accessible olive oil variety. Although the pomace olive oil is prepared, it is still investigated aim proved substitute to other oils in the market and hence is in high requirement as in comparison to the other different types of olive oils. However A deterrent to greater consumption has been its price - an imported one liter bottle of extra virgin can cost between Rs 800 and Rs 1000. A domestically produced one can be cheaper.

25. Medical authorities' recommendations

Olive oil possesses lot of medicinal properties. The high amount of monounsaturated fatty acids (MUFAs), particularly oleic acid, it plays vital role in nutrition, and contains high concentrations of polyphenols, tocopherols and phytosterols, it is important antioxidants and olive oil has been used as important nutrients in the Mediterranean diet. Olive oil prevents neurodegenerative disorders (Parkinson's disease and Alzheimer's disease) and also acts as a cardiovascular and cerebral-vascular protector, due to these medicinal properties many neutraceutical industry has been working with olive oil. All these characteristics make EVOO an essential food component for health. EVOO, also vernacularly known as "liquid gold" is a natural product of inarguable value and not only in the financial sense of the term but also for its renowned properties and advantages on health. It is one of the recommended foods for its nutritive and is essentially a required of the MED, given that to the aspects connected with it.

Based on the chemical structures of EVOO, its coming under the lipophilic products since lipids is the main constituents, primarily MUFA, found by PUFA. This lipid fragments is responsible for preventive properties on cardiovascular disease, autoimmune and inflammatory disorders, provide anti-thrombotic and regulation effects of blood pressure although in a smaller quantities, other compounds such as tocopherols or polyphenols are also present, which are combined with the potent antioxidant and inflammatory activity of EVOO, among other features.

Oleic acid is oil that has the capability to reduce the effects of oxidative stress in the human body, the regulation of the levels of LDL-cholesterol without alteration of HDL-cholesterol and it consequently reduces the risk of heart attacks and other cardiac pathologies. Corresponding to the United States Department of Agriculture

(USDA) 1 tbsp, or 13.5 grams (g) of olive oil, contains 119 calories, 13.5 g of fat, among that 1.86 g is saturated fat rest of it unsaturated and also it provide vitamin E 1.9 milligrams (mg) and 8.13 micrograms (mcg) of vitamin K, apart from that in consists calcium and potassium, as well as polyphenols, tocopherols, phytosterols, squalene, and terpenic acids and other antioxidants. In the Spanish pharmacopeia as well as in other European pharmacopeias, a wide assessment is assigned to the treatment with of olive oil, thus it is recommended as in injectables, because of an accurate role in gynecology, in childhood to old ages. The authors of A Practical Guide on Medicinal Plants indicate that olive oil contains very beneficial oleic acid, phitosterina lecithin and enzymes, in addition to a bitter principle and pigments.

26. Conclusion

Government of India is promoting Olive cultivation under the National Mission on Oil seeds and Oil palm. The diverse agro climatic conditions in India, including the poor and marginal soils can be efficiently utilized for olive production. The government of India's ambitious target of doubling farmer's income can be easily achieved by promotion of Olive plantations in the country. Israeli methodology has grant Rajasthan develop into the core for *India's* new-begin love for *olives*. Rajasthan's tryst with olives and its achievement can now motivate further-more states to approve plantation of this tree, particularly on arid regions. ROCL has distributed nearly 4.5 lakh seedlings to Punjab, Himachal Pradesh, Jammu and Kashmir, UP and NE. In 2017–2018 India imported 10914.86 metric tonnes of oil. That could have huge implications on the livelihood of people in the region. *Olea europaea* L. (Olive Tree) oil considered as golden ingredient in the diet by all food and medical expert across the globe, in addition to offering characteristic organoleptic properties, provides substances capable of preventing the appearance or development of diseases of various nature, from heart and circulatory diseases to metabolic disorders, including carcinogenic processes. To add superior plants to the cultivators, the state government, with Israel's help, has set up a centre of greatness at Jaipur's Bassirange.

Author details

Thangamani Dhandapani[1]*, K.B. Sridhar[2] and S. Vimala Devi[3]

1 ICFRE-Institute of Forest Genetics and Tree Breeding, Coimbatore, India

2 ICAR-Central Research Institute for Dry-land Agriculture, Hyderabad, India

3 ICAR-National Bureau of Plant Genetic Resources, New Delhi, India

*Address all correspondence to: thangamanid@icfre.org

References

[1] Bandelj D, Jakse J, Javornik B. Assessment of genetic variability by microsatellite and AFLP markers. Euphytica. 2004;**136**:93-102

[2] Caballero JM, Río DC. The olive world germplasm bank of Spain. Acta Horticulturae (ISHS). 2008;**791**:31-38

[3] Ruguni E, Biasi R, Muleo R. Olive (Oleaeuropea var. sativa) transformation. In: Jain SM, Minocha SC, editors. Molecular Biology Of Woody Plants. Switzerland, AG: Springer Nature; Vol.2. 2000. pp. 245-279

[4] Barranco D, Trujillo I, Rallo P. Are 'Oblonga' and 'Fantoio' the same cultivar? Hort Science. 2000b; **35**:1323-1325

[5] Baldoni L, Pellegrini M, Mencuccini M, Angiolillo A, Mulas M. Genetic relationships among cultivated and wild olives revealed by AFLP markers. Acta Horticulturae. 2000; **521**:275-284

[6] Available from: https://www.oliveoiltimes.com/world/indias-new-olive-oil-sector/26530

[7] Haouane H, El Bakkali A, Moukhli A, Tollon C, Santoni S, Oukabli A, et al. Genetic structure and core collection of the World Olive Germplasm Bank of Marrakech: Towards the optimised management and use of Mediterranean olive genetic resources. Genetica. 2011;**139**(9): 1083-1094

[8] Verma N, Shaheen R, Yadav SK, Singh AK. Olive (Oleaeuropea L.) introduction in India: Issues and prospects. Vegetos. 2012;**25**(2):44-49

[9] Pantera A, Burgess PJ, Losada RM, Moreno G, López-Díaz ML, Corroyer N, et al. Agroforestry for high value tree systems in Europe. Agroforestry Systems. 2018;**92**:945-959

[10] Mantzanas K, Papanastasis V, Pantera A, Papadopoulos A, Burgess AP. System report: Olive agroforestry in Kassandra. Chalkidiki, Greece, Thessaloniki: Aristotle University; 2015. p. 8

[11] Rosati A, Mantovani D. System Report: Intercropping of Olive Orchards In Italy by Council for Research in Agriculture and the Analysis of Agricultural Economics - Research Center for Food and Nutrition, Via Ardeatina, Rome, Italy; 2015. p. 8

[12] Paris P, Camilli F, Rosati A, Mantino A, Mezzalira G, Dalla Valle C, et al. What is the future for agroforestry in Italy? Agroforestry Systems. 2019; **93**:2243-2256

[13] Mantino A, Ragaglini G, Tozzini C, Cappucci A, Mele M, Bonari E. Yield and nutritive value of alfalfa (*Medicago sativa* L.) in an olive (*Oleaeuropaea* L.) alleycropping practice. In: Book of Abstract 3rd European Agroforestry Conference. France: Montpellier; 2016. pp. 403-406

[14] Sofo A, Manfreda S, Dichio B, Fiorentino M, Xiloyannis C. The olive tree: A paradigm for drought tolerance in Mediterranean climates. Hydrology and Earth System Sciences Discussions. 2007;**4**:2811-2835

[15] Daoui K, Fatemi ZE-A. Agroforestry systems in Morocco: The case of olive tree and annual crops association in Saïs Region. In: Science Policy and Politics of Modern Agricultural System. Netherlands: Springer; Chapter 19. 2014. p. 281

[16] Razouk R, Daoui K, Kajji A, Ramdani A, Chergaoui A. Optimization of rain-fed intercropping system based

on olive trees and annual crops in Northern Morocco. Montpellier: Third European Agroforestry Congress; 2016

[17] Temani F, Bouaziz A, Daoui K, Wery J, Barkaoui K. Olive agroforestry can improve land productivity even under low water availability in the South Mediterranean. Agriculture, Ecosystems and Environment. 2021:307

[18] Ponozzo A, Bernazeau B, Desclaux D. Durum wheat in organic olive orchard: Good deal for the farmers. Agroforestry Systems . 2019;**94**:707-717

Chapter 4

Olive Phenomenon from the Mediterranean Diet: Health Promotion via Phytochemicals

Didar Üçüncüoğlu

Abstract

The Mediterranean Diet contains fruits, vegetables, nuts, whole grains, fish and virgin olive oil (VOO) as a key component. It is well explained that those consumption has a number of positive health effects. It has been accepted for a long time that the leading compound in olive was oleic acid as a monounsaturated fatty acid. However, the latter researches were figured out that VOO rich in natural phenolics have multifaceted influence on major diseases including cancer, diabetes, cardiovascular diseases, neurodegenerative disease, and metabolic disorders. Recent medical studies proved that oleocanthal and oleacein, characteristic bioactive biophenol-secoiridoids in VOO, success in the anti-inflammatory and in the antioxidant properties, respectively. It has more recently investigated that oleocanthal and hydroxytyrosol (HT) kills cancer cells (CCs). HT and oleuropein reduces breast cancer and cutaneous melanoma cancer cells both in number and aggressiveness, and inhibits CCs multiplying. It has been declared too many times that nutrition type is the strongest factor can be caused acute and chronic diseases. However, at the same time, nutrition can also prevent some of those heavy symptoms. The main purpose of presented chapter is to meet olive's bioactive molecules and to examine how to improve our health with diet.

Keywords: phytochemicals, health, olive, virgin olive oil, bioactive compounds

1. Introduction

Olea europaea is naturally cultivated in Mediterranean region. It's drupe fruit (olive), flowers & leaves, seeds, fibers and some of by-products (liquid and solid wastes) are used technologically by humans live in there, and also, this is an agricultural & economic culture of Mediterranean populations (Spain, Turkey, Greece, Italy, Morocco, Tunisia, in particular). However, olive oil is also produced in Australia and the USA and basically [1] consumed European Union Countries, United States, Turkey and China (**Figure 1**). The olive oil is the best demonstrated product that examined both its nutritional and medicinal properties. Ancient historical evidence recommended using VOO for stomach health and against to dermatological ulcer.

Nutritional quality of olives depends on various factors. Cultivar, genetic and biologic factors, soil structure, climatic conditions, harvest time, agricultural applications and extraction type are the most effective parameters [2]. These are

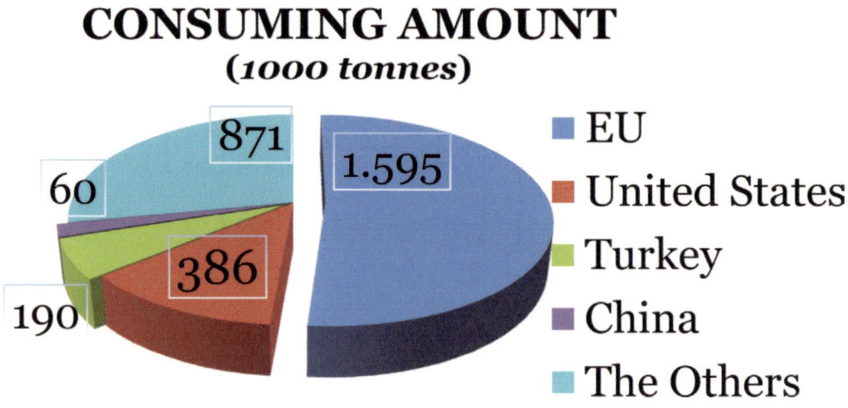

Figure 1.
Leading consumers of olive oil worldwide from in 2020/2021.

also influence on the olive oil quality and its chemical composition. Free acidity determines chemical subtype of olive oils such as lampante, virgin, extra virgin, etc.

The edible oils are commonly well-known due to their unsaturated fatty acids and phenolic profiles. In particular, olive oil is rich in both monounsaturated fatty acids and polyphenols. These attributes basically related to consumption of the Mediterranean Diet. One tablespoon of olive oil contains approximately 120 kcal composed of 10 g monounsaturated lipid. The most important of polyphenols are phenolic acids such as ferulic and vanillic acids, phenolic alcohols like hydroxytyrosol and tyrosol, flavonoids and secoiridoids such as oleuropein [3].

Olive oil is contains triacylglycerols as major components (98–99%) and contains small quantities of free fatty acids, glycerol, phosphatides, pigments, flavor compounds, and sterols as minor components (1–2%). Approximate fatty acid composition of olive oil as fallows: 14–15% saturated fats (13% palmitic acid-C16:0 and 1–2% stearic acid-C18:0) and 85% unsaturated fats (60–70% oleic acid-C18:1, also known as ω-9 and 12–15% linoleic acid-C18:2, also known as ω-6 fatty acid, and 3% palmitoleic acid). This profile shows us olive oil composed of more monounsaturated than polyunsaturated fatty acids and it is free of trans (E-) form of fatty acids. Previous researches indicated that a higher proportion of monounsaturated fatty acids in diet cause a strong reduction in the cardiovascular disease (CVD) risks.

The polyphenols are natural antioxidants that link oil's bitter taste and astringency. Moreover, they raise the resistance capacity against to oxidation. The bioactive compounds namely phenolic antioxidants (oleic acid, squalene, terpenoids), simple phenols (hydroxytyrosol, tyrosol), secoiridoids (oleuropein, aglycone of ligstroside, and their respective decarboxylated dialdehyde derivatives, oleacein, oleochantal) and lignin, vitamin E contribute to oil different biologic character such as antitumor, antimicrobial and antioxidant effects via possible pathophysiological pathways generally associated with a decreased risk in breast, colon and skin cancer. It needs to be noted that the quantity of these bioactive molecules depends on many factors such as olive variety, environmental factors (altitude, cultivation practices), harvest time (unripe, semi, ripe olives the extraction and storage conditions [4].

Here, in this chapter, phytoactive compounds found in olive are determined chemically and then biologic mechanisms are described how they prevent us some chronic diseases.

2. Phytochemicals

2.1 Phenolics and tocopherols

The presences of phenolics and tocopherols affect radical scavenging activity and so, antioxidant capacity of olive oil. These chemicals also considered as a functional food component, contains different classes of phenols: simple phenols, phenolic acids, phenolic alcohols, secoiridoids, lignans, and flavonoids. According to EFSA [5], molecules from olive (hydroxytyrosyl and oleuropein complex) have beneficial effects on human health. The consumption of this complex approximately 2–15 mg/day is essential. The vitamin E is a fat-soluble vitamin and is composed by α-, β-, γ and δ-tocopherols and α-, β-, γ- and δ-tocotrienol isomers. α-tocopherol has the highest antioxidant activity in fat-riched foods. Moreover, vitamin E and phenol mixtures are very promising oil stabilizing agent during frying, thanks to its capacity to reduce acrolein and acrylamide production. According to EFSA's current scientific opinion [6] an average requirements of α-tocopherol could be 11–13 mg/day for adults. Extra virgin olive oil contains 100–760 mg/kg α-tocopherol [7, 8].

Mediterranean country peoples demonstrate a long life and lower case of the age-related diseases due to their food habits. Thus, the consumption of extra virgin olive oil has been clinically and experimentally associated with the health promoting properties of the Mediterranean diet. Extra virgin olive oil is characterized by a high amount of polyphenols as miracle pharmanutritional properties against to the development of several chronic diseases such as cardiovascular diseases, diabetes, obesity, metabolic syndrome and cancer. Squalene is a hydrocarbon composed in olive oil and acts as a weak antioxidant affect. Especially, its importance comes from the protective effect under heating against to the oxidative stress. Besides, the squalene quantity varies between 2 and 7 g/kg of olive oil. Scientific papers prove us the health benefits of squalene and its influence capability on cancer cells [9–11]. Meanwhile, tocopherols have attracted tremendous attentions because of their potential roles in preventing aging-associated diseases such as cardiovascular and Alzheimer's disease [12–14]. Oleacein and oleochantal (**Figure 2**) are among the major phenolics determined in virgin olive oil. These are the dialdehydic forms of the elenolic acid deriving from oleuropein and ligstroside, respectively. They have positive role in the treatment of cardiovascular pathologies, certain types of cancers, chronic inflammatory diseases, Alzheimer's disease, Helicobacter pylori infection [14]. Cutaneous melanoma is a type of cancer that is spreading in Europe. The relation between melanoma risk and consumption of olive oil has been recently studied. Oleuropein, the main secoiridoid glucoside present in the *Olea europaea* leaves have already been investigated in melanoma *in vitro* and *in vivo* models revealing their cytotoxic activity and anticarcinogenic action. Oleacein, another abundant secoiridoid in olive and leaves show antitumor activity via controlling the altered pattern of gene expression (transcripts and miRNAs) related to mTOR and BCL2 pathways in melanoma cells. Hydroxytyrosol, the most representative simple phenol of *Olea europaea* L. leaves and virgin oils, causes inhibition of melanoma cell proliferation activating caspase-3-dependent apoptosis [15, 16].

2.2 Phytosterols

Phytosterols, in other words plant sterols, are bioactive compounds of seeds, fruits, cereals, and nuts. They can be found in free form, esterified with fatty acids or conjugated with glycosides. Campesterol, β-sitosterol and stigmasterol are the most common phytosterols. Phytosterols are by-products of isoprenoid biosynthesis

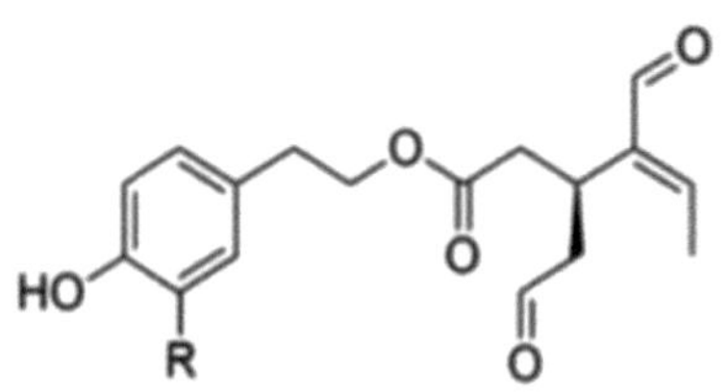

Figure 2.
Chemical structure of Oleacein and Oleochantal.

pathway via squalene from acetyl coenzyme A. Its generation cascade involving more than 30 enzymes catalyzed reactions in similar cholesterol biosynthesis within cell membranes. All phytosterols contain one double bond at carbon-5 position and saturation of this double bond occurs either enzymatically in vivo or through hydrogenation. Both β-sitostanol and campestanol are the two most common stanols. In olive oil, they are esterified with fatty acids like esters of sitostanol (sitostanyl oleate) and esters of campestanol (campesteryl oleate). Human cells cannot synthesize those endogenously, so we have to obtain them from our diet. The daily intake of plant stanols (saturated sterols) is about 25 mg/day compared to sterols ranging from 150 to 400 mg/day which include 65% of intake as β-sitosterol, 30% as campesterol and 5% as stigmasterol. The maximum values of campesterol, stigmasterol and β-sitosterol were found in different types of olive oil 34.46 mg, 29.56 mg and 259.46 mg/100 g oil, respectively. Serum phytosterol level in humans should range from between 2.9 and 17.0 mg/L. Phytosterols are alcohols for organic chemistry, resemblance to cholesterol both in structure and in function. Their function in plant cells is to regulate the fluidity and permeability of cell membranes in a way similar to cholesterol in cell membranes of mammals. They can inhibit the absorption of cholesterol by its direct displacement from micelles. This reported as the cholesterol-lowering effect of phytosterols in humans. Additionally, phytosterols and their derivatives have high health benefits related to antiinflammatory, antiulcerative, antitumor, antibacterial, and cardioprotective properties [17–23].

2.3 Chlorophyll and carotenoids

The minor contents in olive oil give to its characteristic organoleptic and nutritional features. Among them, color pigments play an important role in oxidative stability during storage and in the preservation of its quality because of their antioxidant characteristics and prooxidant roles under light. The pigments used as indicators to indicate olive oil freshness or storing conditions. For example, low amount of pheophytins show a fresh, well-stored olive oil, while high presence of those indicate an old, bad-stored oil [24, 25]. Pigments are divided into two main groups in foods namely carotenoids and chlorophylls. Chlorophylls are responsible for the greenish color of extra virgin olive oil. On the other side, β-carotene, lutein, zeaxanthin and xanthophylls called carotenoids are natural antioxidants found in various amount in olive oil. They have beneficial properties for human health. β-carotene appears to lower the risk of heart diseases, while lutein protects the eye retina from oxidative damage. Some other studies were showed that carotenoids exhibit bioactivities in photosynthesis, special antioxidant properties and produce improvements in cognitive function and cardiovascular health [26–30].

3. Clinical and experimental studies on chronic diseases

The potential health advantages caused by olive oil consumption, particularly within the context of the Mediterranean diet, have been extensively investigated. As mentioned before, olive oil consumption was found to be beneficial for some of chronic non communicable diseases [31]. Below-explained clinical results are also given in **Table 1** with references.

Scope	Outcomes	References
Olive oil consumption and cancer risk	It can decrease the risk of upper digestive and respiratory tract neoplasms, breast and colorectal cancer sites	Pelluchi et al. [32]
Whether olive oil or MUFA intake was associated with the development of cancer	Olive oil intake has a protective role on cancer risk. But, it was not clear which component is responsible for the beneficial effects	Psaltopoulou et al. [33]
MUFA effect on CVD and mortality	VOO reduced risk of all-cause mortality, CV mortality, stroke	Schwingshackl Hoffman [34]
VOO intake and CHD & stroke risks correlation	VOO intake had a significant protective effect	Martinez-Gonzalez et al. [35]
Determine inflammation or endothelial effect of olive oil	Olive oil might be exert benefical effects on endothelial function and markers of inflammation and endothelial function	Schwingshackl et al. [36]
VOO with high phenolic content, CVD effect and Nutraceutical prevention	Provides small benefical effect on systolic blood pressure and serum oxidative status. Considered as nutraceutical in CV prevention	Hohmann et al. [37]
Breast cancer risk and dose–response	Olive oil might be a protective factor for the cancer occurrence among case–control studies	Xin et al. [38]
To compare olive and other edible oils effects on blood lipid level	VOO intake decreased serum total cholesterol, LDL and triglycerides less but increased HDL more than other plant oils	Ghobadi et al. [39]
High versus low polyphenol content on CVD risk factors	olive oils contains high polyphenol show some CVD risk reductions	George et al. [40]
To illustrate the impact of olive oil on type II diabetes	Olive oil intake could be benefical for the prevention and management of type II diabetes	Schwingshackl et al. [41]
to investigate effects of 2 oil blends phytonutrient concentrations on blood lipid profile, compared with refined olive oil as a control	borderline hypercholesterolemia extent as refined olive oil	Haldar et al., [42]

Table 1.
Experimental and clinical studies about olive, nutrition and health [32–42].

3.1 Cardiovascular diseases (CVD)

A comprehensive work included 101,460 cases of Coronary Heart Disease (CHD) and 38,673 cases of stroke participants revealed that after every 25 g increase in olive oil consumption, the risk of CHD was reduced by approximately 4%, while the risk stroke was decrease by 26%. Another research aimed to assess the consumption of monounsaturated fatty acids and olive oil intake on human mortality was applied 32 sample group contains total 841,211 cases. Overall, higher oil intakes lead to a lower risk for all-cause mortality, cardiovascular mortality, cardiovascular events and stroke. On the other hands, the other lipid origin such from animal had no significant positive effect on morbidity and mortality in this work. A research including 3106 individuals search try to detect the potential mediators of the olive oil/CVD relationship. The daily consumption between 1 mg and 50 mg resulted in a significantly more pronounced decrease in C-reactive protein and interleukin-6. The other study examined the effect of high versus low concentration of polyphenol in olive oil on CVD risk factors in clinical trials. It was found that high polyphenol olive oil has beneficial effects on malondialdehyde, oxidized-LDL, total cholesterol and HDL cholesterol, suggesting that olive oil may have cardioprotective properties. A comparative research, which aimed to explain the effects of olive oil or some of other edible oils consumption on blood lipids, showed that olive oil intake significantly diminish total cholesterol, LDL and triglyceride less, but increased HDL more than the others. Chronic kidney disease (CKD) is one of the most common chronic degenerative diseases and it carries a considerable risk factor for CV disorders and mortality. Noce et al., [43] study shows that daily use of high phenolic contents from extra virgin olive oil contributes anti-inflammatory and antioxidant action in nephropathic patients. Consequently, the essential role has been highlighted again here that a well-balanced nutritional-diet therapy plays in the clinical management of CKD cases and the eventual improvements in patient's life style, performing positive effects on CKD symptoms.

3.2 Cancer

There are too actual clinical meta-analysis researches. They are mainly focused between olive oil consumption and various cancer types. 37,140 participants from 19 case–control studies were revealed that extra virgin olive oil intake was associated with low probability to have any type of cancer, compared with the lowest quality of olive oil intake. The other study evaluated the effect of olive oil consumption on breast cancer risk. The results stated that olive oil can decrease the risk of breast cancer as well as other types of cancer. Another comprehensive research, included 150,000 females from 5 cohort and 11 retrospective case control studies, demonstrated that higher olive oils consumption showed a protective effect against breast cancer. Some clinical studies show that the mixtures of phytosterols have inhibitory effect on 5α-reductase, give rise to apoptosis, alter testosterone metabolism, change urine flow and prostate volume. On the other hand some other experimental studies explain that phytosterols prevent lipoprotein oxidation, inhibit cell growth and the expression of caveolin-1, decrease cell proliferation, reduce or inhibit hyper proliferation of colonocytes, reduce in tumor size. These are the various mechanisms to express anticancer features from phytosterols of olives and olive oils [19, 44]. More recently, Bartolomei [45] provides a huge dataset to characterize of olive oils antioxidant properties, both in vitro by Trolox equivalent antioxidant capacity, oxygen radical absorbance capacity (ORAC value), ferric reducing antioxidant power (FRAP value), and 2,2-diphenyl-1-picrylhydrazyl assays, and at the cellular level in hepatic (HepG2) and intestinal (Caco-2) cells. One of the highlighted results of this

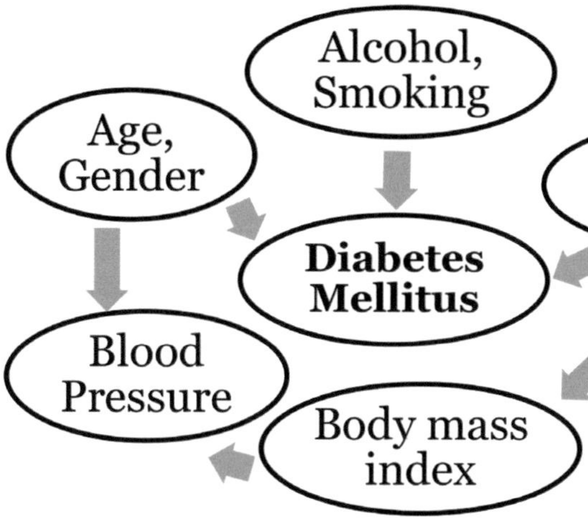

Figure 3.
Risk factors in diabetes mellitus.

study was to determine the antioxidant effects of phenolic compounds in an extra virgin olive oil (EVOO) extract, using either in vitro assays or liver and intestinal cell models, rather than the effects of single phenols, such as hydroxytyrosol or oleuropein. The selective trans-epithelial transport of some oleuropein derivatives was observed for the first time in differentiated Caco-2 cells in this research.

3.3 Type II diabetes

Risked factors for diabetes Mellitus (DM) has been described [46] as fallows: physical inactivity, first degree relative with DM, members of high-risk ethnic populations (such as African Americans, Latinos, Native Americans, Asian American and Pacific Islanders), women who delivered baby, hypertension, HDL cholesterol <35 mg/dL, triglycerides >250 mg/dL, women with polycystic ovary sendrome, HbA1C ≥ 5.7%, impaired glucose tolerance, ımpaired fasting glucose, obesity, insulin resistance and finally history of CVD (**Figure 3**). It was proved that 10 g daily increasing intake of olive oil was caused 9% reduced risk of type-II diabetes, while when the highest quality olive oil intake category was compared to the lowest quality intake category, it was revealed a 16% reduced risk of diabetes. This meta-analysis collected data from 187,068 individuals, participating in 4 cohort studies and 29 trials. Therefore, this meta-analysis provides beneficial evidence concerning the risk of type II diabetes consuming olive oil [42].

4. Conclusion

In biological systems, free radical reactions are associated with aging, cancer, cardiovascular disease, optical disease, and neurodegenerative disease. Antioxidants are used to prevent such reactions and are therefore supporting health. Regular consumption of olive oil has a strong correlation with prevention of cardiovascular disease, immune function, and gastrointestinal (GI) disease. Olive oil improves the curing of mortal diseases like cancer and liver disease. Dietary use of olive oil is therefore helpful in maintaining good health. The present chapter evaluating the effect of olive oil consumption on human health, identified the protective effects of olive oil on all-cause and cardiovascular mortality, as well as on cardiovascular events. Olive oil consumption exhibits a protective role against overall and particularly breast cancer occurrence, as well as diabetes mellitus type 2. In general, benefits which have been recorded from the intake of pure olive oil have also been recorded from the adoption of a health-promoting lifestyle, based on the principles of the Mediterranean diet. Diets including plants foods, fish, and seafood and preferably including plant oils and low-fat dairy products are associated with a lower risk of most chronic diseases. In conclusion, the aggregated evidence corroborates that the specific impact of olive oil consumption is beneficial for human health, and particularly for the prevention of cardiovascular diseases, several cancers, and diabetes mellitus.

Conflict of interest

The author declares no conflict of interest.

Author details

Didar Üçüncüoğlu
Faculty of Engineering, Department of Food Engineering, Çankırı Karatekin University, Çankırı, Turkey

*Address all correspondence to: didaru@karatekin.edu.tr

References

[1] International Olive Council (IOC), 2021. Retrieved from https://www.statista.com/statistics/940532/olive-oil-consumption-worldwide-by-leading-country/ (Accessed in 26.02.2021)

[2] Martakos I, Kostakis M, Dasenaki M, Pentogennis M, Thomaidis N. Simultaneous Determination of Pigments, Tocopherols, and Squalene in Greek Olive Oils: A Study of the Influence of Cultivation and Oil-Production Parameters. Foods. 2019;9(1):31. DOI:10.3390/foods9010031

[3] Zeb A, Murkovic M. Olive (*Olea europapea* L.) Seeds, From Chemistry to Health Benefits. Chapter 100 - Nuts and Seeds in Health and Disease Prevention 2011; 847-853, Academic Press. DOI: 10.1016/B978-0-12-375688-6.10100-8

[4] Vita di S. Extra virgin olive oils storage: Effect on constituents of biological significance, Chapter 24. Editor(s): Victor R. Preedy, Ronald Ross Watson, Olives and Olive Oil in Health and Disease Prevention, Academic Press. 2021;291-297, ISBN 9780128195284. DOI:10.1016/B978-0-12-819528-4.00029-8.

[5] EFSA Panel on Dietetic Products, N. a. A. N. (2011) Scientific Opinion on the substantiation of health claims related to oleic acid intended to replace saturated fatty acids (SFAs) in foods or diets and maintenance of normal blood LDL-cholesterol concentrations (ID 673, 728, 729, 1302, 4334) and maintenance of normal (fasting) blood concentrations of triglycerides (ID 673, 4334) pursuant to Article 13(1) of Regulation (EC) No 1924/2006. EFSA Journal 9, 2043.

[6] EFSA Panel on Dietetic Products, N. a. A. N. (2015) Scientific opinion on the substantiation of health claims related to polyphenols in olive and protection of LDL particles from oxidative damage (ID 1333, 1638, 1639, 1696, 2865), maintenance of normal blood HDL-cholesterol concentrations (ID 1639), maintenance of normal blood pressure (ID 3781), "anti-inflammatory properties" (ID 1882), "contributes to the upper respiratory tract health" (ID 3468), "can help to maintain a normal function of gastrointestinal tract" (3779), and "contributes to body defences against external agents" (ID 3467) pursuant to Article 13(1) of Regulation (EC) No 1924/20061. EFSA J.1 9, 2033.

[7] Lucci P, Bertoz V, Pacetti D, Moret S, Conte L. Effect of the Refining Process on Total Hydroxytyrosol, Tyrosol, and Tocopherol Contents of Olive Oil. Foods. 2020; 9:292. DOI:10.3390/foods9030292

[8] Dugo L, Russo M, Cacciola F, Mandolfino F, et al. Determination of the Phenol and Tocopherol Content in Italian High-Quality Extra-Virgin Olive Oils by Using LC-MS and Multivariate Data Analysis. Food Analytical Methods. 2020; 13:1027-1041. DOI: 10.1007/s12161-020-01721-7

[9] Newmark HL. Squalene, olive oil, and cancer risk: A review and hypothesis.Cancer Epidemiol. Biomark. Prev.1997,6, 1101-1103

[10] Rodriguez-Rodriguez, R.; Simonsen, U. Natural Triterpenoids from Olive Oil: Potential Activities AgainstCancer. InNatural Compounds as Inducers of Cell Death; Diederich, M., Noworyta, K., Eds.; Springer: Cham, Switzerland, 2012; Volume 1, pp. 447-461.

[11] Chira M, Rigane G, Calas-Blanchard C et al. Phenolic, tocopherols and squalene profiles (HPLC-UV) of chemlali-sfax olive oil according to extraction procedure. Revue Roumaine

de Chimie 2020; 65:2, 179-190 DOI: 10.33224/rrch.2020.65.2.07

[12] Azzi A. Many tocopherols, one vitamin E. Molecular Aspects of Medicine. 2018;61:92-103. DOI: 10.1016/j.mam.2017.06.004

[13] Zhang L, Wang S, Yang R, et al. Simultaneous determination of tocopherols, carotenoids and phytosterols in edible vegetable oil by ultrasound-assisted saponification, LLE and LC-MS/MS. Food Chemistry 289 (2019) 313-319. DOI: 10.1016/j.foodchem.2019.03.067

[14] Macciola V, Cuomo F, de Leonardis A. IMPORTANCE OF OLEacein and oleochantal on the oxidative stability of extra virgin olive oil measured by Rancimat. La rivista italiana delle sostanze grasse. 2020; XCVII. https://www.innovhub-ssi.it/kdocs/1990259/2020_Vol._97_-_Art._03_Macciola.pdfRuzzolini, 2018;

[15] Cirmi S, Celano M, Lombardo GE, Maggisano V, Procopio A, Russo D et al. Oleacein inhibits STAT3, activates the apoptotic machinery, and exerts anti-metastatic effects in the SH-SY5Y human neuroblastoma cells. Food Funct. 2020;11 (4), 3271-3279. DOI: 10.1039/d0fo00089b

[16] Carpi S, Polini B, Manera C, Digiacomo M, Salsano JE, Macchia M, Scoditti E, Nieri P. miRNA Modulation and Antitumor Activity by the Extra Virgin Olive Oil Polyphenol Oleacein in Human Melanoma Cells. Frontiers in Pharmacology 2020;11. DOI:10.3389/fphar.2020.574317

[17] Uddin MS, Sarker MZI, Ferdosh S, et al. Phytosterols and their extraction from various plant matrices using supercritical carbon dioxide: a review, J. Sci. Food Agric. 2015;95:7, 1385-1394.

[18] Ogbe RJ, Ochalefu DO, Mafulul SG, Olumide OB. A review on dietaryphytosterols: their occurrence, metabolism and health benefits, Asian J. Plant Sci. Res. 2015;5:4, 10-21.

[19] Shahzad N, Khan W, Shadab MD, Ali Asgar et al. Phytosterols as a natural anticancer agent: Current status and future perspective. Biomedicine & Pharmacotherapy 88 (2017) 786-794. DOI: 10.1016/j.biopha.2017.01.068

[20] de Figueiredo LC, Bonafe EG, Martins JG, Martins, AF et al. Development of an ultrasound assisted method for determination of phytosterols in vegetable oil. Food Chemistry. 2018; 240:441-447. DOI:10.1016/j.foodchem.2017.07.140

[21] Huang J, Xu M, Fang YJ, Lu MS, Pan ZZ, Huang WQ, Chen YM, Zhang CX, Association between phytosterol intake and colorectal cancer risk: acase-control study, Br. J. Nutr. 2017; 117:6, 839-850.

[22] Wang H, Liu H, Guo B, Lan D et al. A novel poly (NMA-co-DEA-co-EDMA) monolithic column as a sorbent for online solid-phase extraction and its application in the determination of beta-sitosterol in plant oil samples, Food Chem. 2019; 278:594-600.

[23] Silva ACA, Baggio SR, Mariutti LRB, Bragagnola N. One-step rapid extraction of phytosterols from vegetable oils. Food Research International 2020;130:108891 DOI:10.1016/j.foodres.2019.108891

[24] Giuliani A, Cerretani L, Cichelli A. Chlorophylls in olive and in olive oil: Chemistry and occurrences. Crit. Rev. Food Sci. Nutr.2011;51, 678-690.

[25] Cayuela JA, Yousfi K, Carmen Martínez M, García, JM. Rapid Determination of Olive Oil Chlorophylls and Carotenoids by Using Visible Spectroscopy.J. Am. Oil Chem. Soc.2014;91, 1677-1684

[26] Tapiero H, Townsend DM, Tew KD. The role of carotenoids in the prevention of human pathologies. Biomed Pharm.2004;58, 100-110.

[27] Fieor J, Bra K. Potential role of carotenoids as antioxidants in human health and disease. Nutrients, 2014;6, 466-488.

[28] Hashimoto H, Sugisaki M, Yoshizawa M. Ultrafast time resolved vibrational spectroscopies of carotenoids in photosynthesis. Biochimica et Biophysica Acta (BBA) – Bioenergetics. 2015;1847(1), 69-78. DOI: 10.1016/j.bbabio.2014.09.001

[29] Egeland ES. Carotenoids. In The Physiology of Microalgae; Borowitzka M, Beardall J, Raven J. Developments in Applied Phycology; Springer International Publishing: Cham, Switzerland, 2016;6.

[30] Eggersdorfer M, Wyss A. Carotenoids in human nutrition and health. Archives of Biochemistry and Biophysics, 2018; 652, 18-26.

[31] Foscolou A, Critselis E, Panagiotakos D. Olive oil consumption and human health: A narrative review. Maturitas 2018;11860:66. DOI:10.1016/j.maturitas.2018.10.013

[32] Pelucchi C, Bosetti C, Negri E, Lipworth L, La Vecchia C. Olive oil and Cancer risk: an update of epidemiological findings through 2010. Curr. Pharm. Des. 2011; 17, 8: 805-812, DOI: 10.2174/138161211795428920

[33] Psaltopoulou T, Kosti RI, Haidopoulos D, Dimopoulos M, Panagiotakos DB. Olive oil intake is inversely related to cancer prevalence: a systematic review and a meta-analysis of 13800 patients and 23340 controls in 19 observational studies. Lipids Health Dis. 10. 2011; 127. DOI: 10.1186/1476-511X-10-127

[34] Schwingshackl L, Hoffmann G. Monounsaturated fatty acids, olive oil and health status: A systematic review and meta-analysis of cohort studies, Lipids Health Dis. 2014; 13:154. DOI: 10.1186/1476-511X-13-154

[35] Martínez-González MA, Dominguez LJ, Delgado-Rodríguez M. Olive oil consumption and risk of CHD and/or stroke: a meta-analysis of case-control, cohort and intervention studies. Br. J. Nutr. 2014; 112, 2:248-259. DOI: 10.1017/S0007114514000713

[36] Schwingshackl L, Christoph M, Hoffmann G. Effects of olive oil on markers of inflammation and endothelial function-a systematic review and meta-analysis. Nutrients 2015; 7, 9: 7651-7675, DOI: 10.3390/NU7095356

[37] Hohmann CD, Cramer H, Michalsen A, Kessler C, Steckhan N, Choi K, Dobos G. Effects of high phenolic olive oil on cardiovascular risk factors: a systematic review and meta-analysis, Phytomedicine 2015; 22, 6:631-640, DOI: 10.1016/j.phymed.2015.03.019

[38] Xin Y, Li XY, Sun SR, Wang LX, Huang T. Vegetable oil intake and breast cancer risk: a meta-analysis. Asian Pac. J. Cancer Prev. 2015;16, 12: 5125-5135. DOI: 10.7314/apjcp.2015.16.12.5125

[39] Ghobadi S, Hassanzadeh-Rostami Z, Mohammadian F, Nikfetrat A, negar H Ghasemifard, Raeisi Dehkordi, Faghih S. Comparison of blood lipid-lowering effects of olive oil and other plant oils: a systematic review and meta-analysis of 27 randomized placebo-controlled clinical trials. Crit. Rev. Food Sci. Nutr. 2018; 8: 1-15. DOI: 10.1080/10408398.2018.1438349

[40] George ES, Marshall S, Mayr HL, Trakman GL, Tatucu-Babet OA, Lassemillante A-CM, Bramley A,

Reddy AJ, Forsyth A, Tierney AC, Thomas CJ, Itsiopoulos C, Marx W. The effect of high-polyphenol extra virgin olive oil on cardiovascular risk factors: a systematic review and meta-analysis, Crit. Rev. Food Sci. Nutr. 2018; 30, 1-138, DOI: 10.1080/10408398.2018.1470491

[41] Schwingshackl L, Lampousi AM, Portillo MP, Romaguera D, Hoffmann G, Boeing H. Olive oil in the prevention and management of type 2 diabetes mellitus: a systematic review and meta-analysis of cohort studies and intervention trials, Nutr. Diabetes 2017; 7, 4:e262. DOI: 10.1038/nutd.2017.12

[42] Haldar S, Long Hui Wong, Shia Lyn Tay, Jörg J Jacoby, Pengfei He, Farhana Osman, Shalini Ponnalagu, Yuan Rong Jiang, Hwee Peng Rebecca Lian, Christiani Jeyakumar Henry, Two Blends of Refined Rice Bran, Flaxseed, and Sesame Seed Oils Affect the Blood Lipid Profile of Chinese Adults with Borderline Hypercholesterolemia to a Similar Extent as Refined Olive Oil, The Journal of Nutrition, 2020;150:12, 3141-3151, DOI:10.1093/jn/nxaa274

[43] Noce A, Marrone G, Urciuoli S, Di Daniele F, Di Lauro M, Pietroboni Zaitseva A, Di Daniele N, Romani A. Usefulness of Extra Virgin Olive Oil Minor Polar Compounds in the Management of Chronic Kidney Disease Patients. Nutrients. 2021; 13(2):581. DOI:10.3390/nu13020581

[44] Stark AH, Madar Z, Olive oil in the prevention of breast and colon carcinogenesis, Chapter 28. Editor(s): Victor R. Preedy, Ronald Ross Watson, Olives and Olive Oil in Health and Disease Prevention, Academic Press. 2021;337-345. ISBN 9780128195284. DOI: 10.1016/B978-0-12-819528-4.00047-X.

[45] Bartolomei M, Bollati C, Bellumori M, Cecchi L, Cruz-Chamorro I, Santos-Sánchez G, Ranaldi G, Ferruzza S, Sambuy Y, Arnoldi A, Mulinacci N, Lammi C. Extra Virgin Olive Oil Phenolic Extract on Human Hepatic HepG2 and Intestinal Caco-2 Cells: Assessment of the Antioxidant Activity and Intestinal Trans-Epithelial Transport. Antioxidants. 2021; 0(1):118. DOI:10.3390/antiox10010118

[46] Msopa E, Mwanakasale V. Identification of risk factors of diabetes mellitus in bank employees of selected banks in Ndola town. Diabetes & Metabolic Syndrome: Clinical Research & Reviews. ISSN 1871-4021, 2019; 13, Issue 2: 1497-1504. DOI:10.1016/j.dsx.2018.11.062

Chapter 5

The Impact of Olive Oil and Mediterranean Diet on the Prevention of Cardiovascular Diseases

Muhammad Akram, Rumaisa Ansari, Naheed Akhter, Olutosin Ademola Otekunrin, Sadia Zafar, Muhammad Ishaque, Naveed Munir, Luigi Sciarra, Giulia My, E.C. Gianvito Matarrese, Zefferino Palamà and Muhammad Riaz

Abstract

The Mediterranean diet has a lot of health benefits but especially because it lowers the incidence of cardiovascular diseases. It has been shown that food components, certain nutrients and the pattern of the diet lowers the risk of several diseases such as diabetes, certain cancers, obesity, respiratory disorders, mental health and cognitive decline, bone diseases (osteoarthritis), healthy aging and quality of life among more others. It has been concluded from studying the mechanism responsible for lowering these risks that food combinations, food nutrients, presence of non-nutritive substances, lifestyles habits and the cooking techniques all together make the Mediterranean dietary pattern into a tool that can not only prevent but can also be used as a way of treatment for these medical ailments. As part of the essential dietary fat, consumption of extra virgin olive oil is the main feature of Mediterranean diet. Olive oil is noted to have anti-bacterial characteristics, involved in improving the endothelial function in young females, and is hypothesized to have epigenetic effects interplay offering protection from cancers due to the presence of beneficial monounsaturated fats. The presence of antioxidants contributes to the inflammation protecting properties of the olive oil. Olive oil has high quantities of antioxidants and offers numerous benefits for cardiovascular health, such as protection of LDL from oxidation and lowering of the high blood pressure as well as offers protection from diabetes mellitus. The Mediterranean diet and the Olive oil consumption also have a fundamental impact in secondary prevention, such as in patients with atrial fibrillation that underwent catheter ablation.

Keywords: Mediterranean diet, olive oil, cardiovascular diseases, benefits of olive oil consumption, atrial fibrillation

1. Introduction

Native to the Mediterranean region, olive tree also known as *Olea europaea* is considered to be one of the oldest trees still present (**Figure 1**). The fruits and products obtained from the olive tree such as olive oil have been used for a long time for the diet and nutritional demands of the residing population in the Mediterranean regions. As part of the Mediterranean culture olive trees have been cultivated and harvested for a long time. Olives are taken and subsequently used for olive oil extraction. The importance of the use of olive oil in both daily and cultural practices is evident by the fact that the Greek philosophers began to search on its medicinal and nutritional properties by as early as the seventh century BC. Particularly ingestion of olive oil and its use as an ointment for treating ailments related to dermatological ulcers and stomach was recommended by both Hippocrates and Aristotle who recognized the health benefits attributed to the olive oil consumption [1]. Mediterranean countries such as Turkey, Spain, Greece, Morocco, Italy and Tunisia are involved in most (70%) of the olive oil production [2]. Apart from these some other countries such as USA and Australia are also involved in the production of olive oil. Different types of olive oil with different profiles of polyphenol content are being produced by a single country [3] showing that olive oil of different varieties and qualities exists. Nutritional characteristics of different types of olive oil also differ.

Animal fats mainly constitute the saturated fatty acids, while the fats of vegetables mostly contain polyunsaturated fatty acids. Especially, olive oil obtained from the fruits of olive tree consists of both the polyphenols and monounsaturated fatty acids. Even today, the health benefits of consuming olive oil in place of fats of other origins remain unexcelled and is most likely attributed to the extraordinary chemical composition of olive oil [4]. However, the benefits of the olive oil for human health have been mostly evaluated in attribution to the consumption of the Mediterranean diet [5], whereas its individual beneficial impact on human health remains to be confirmed by more research [6].

Figure 1.
Millennial olive tree in Italy, Puglia, San Vito dei Normanni (Mrs. Teresa Village).

There are different types of the marketed olive oil with difference in their potential healthful impacts and include pomace oil, refined olive oil, virgin olive oil and extra virgin olive oil etc. [7]. In context of research the different types of olive oil are often not individually evaluated for their potential health benefits and therefore the difference in their beneficial properties is not properly evaluated. Extra virgin olive oil is the most naturally made olive oil and is produced within the first 24 hours after the harvest and from the very first pressing of the olives. Extra virgin olive oil is obtained by the mechanical means rather than the chemical and no excessive heating is underwent during its extraction as the temperature for extraction is fixed at degrees lesser than 28°C. The acidity or free fatty acid level of the oil is lesser than 0.8 percent contributing to the oil's optimal odor and taste. It should be noted that olive fruit contains different types of hydrophilic phenolic compounds such as simple phenolic compounds, complex phenolic compounds and lignans. Simple phenolic compounds include vanillic, coumaric, galls, caffeic acids, tyrosol and hydroxytyrosol etc. whereas complex phenolic compounds contain secoiridoids such as ligstroside and oleuropein. Lignans found in the olives include pinoresinol and 1-acetoxypinoresinol etc. Extra virgin olive oil has the highest amounts of phenolic compounds as compared to the virgin olive oil [8]. Lignans on the other hand, are not found in any refined olive oils but only found in extra virgin olive oil [9]. Because of these extraordinary compositions extra virgin olive oil is the most beneficial to human health as compared to other refined oils and is used as a precautionary tool against the cardiovascular diseases [10]. Virgin olive oil just like the extra virgin olive oil also arises from the very first pressing of the fruits of olive tree but the acidity levels are a little higher in comparison i.e. lesser than 2 percent. Virgin oil contains in its composition the hydrophilic phenols such as phenolic acids alcohol, lignans, flavonoids and secoiridoids [11]. But because these compounds are present in comparatively lower levels, its consumption is not as healthful as compared to the extra virgin olive oil. But the consumption of virgin olive oil as compared to the other refined oils is still recommended [12]. Another types of marketed olive oil is the refined olive oil which is refined by chemical and thermal means in order to get as much oil from the pulp of olives as possible. This pulp is the leftover of the first pressing of olives. Acids and alkalis are the chemical agents used for the purpose of refinement. This process of oil extraction results in oil with higher fat and acidic content as compared to the extra virgin and virgin olive oil. Therefore, olive oil obtained and refined this way lacks the optimal odor, taste and natural antioxidants. As compared to the olive oils obtained from the first pressing of olives, the refined olive oil has lesser antioxidant potential. Moreover, p-hydroxyphenylacetic acid, p-coumaric acid, o-coumaric acid, vanillic acid and p-hydroxybenzoic acid present in the refined olive oil offer very little protection and no antioxidant potential [13]. Pomace oil is another type of the marketed olive oil and is produced as a by-product in the production of extra virgin olive oil. This oil is obtained by heating the olive seeds, skins and pulp. The remaining oil is extracted by the use of hexane as a solvent. The resultant oil obtained is then further refined and therefore it offers very limited beneficial effects on human health as the antioxidant potential is very low. Still more other types of oils produced through more poor practices e.g. glampante oil. These oils, unless refined, are not considered to be suitable for human use [14].

2. Mediterranean diet and cardiovascular diseases

The Mediterranean diet was defined in 1960 by Ancel Keys circa [15] and is one of the most well known and most studied dietary pattern in the world [Medscape,

Figure 2.
Oil-Garlic and Pepper. The application of oil in the kitchen leads to this famous dish in the all the Mediterranean diet. On the right-side pure drop of extra virgin olive oil (Gianvito Matarrese EVO Restourant Alberobello-Italy).

PubMed impact]. The Mediterranean diet has a lot of health benefits but especially because it lowers the incidence of cardiovascular diseases. It has been shown that food components, certain nutrients and the pattern of the diet lowers the risk of several diseases such as diabetes, certain cancers, obesity, respiratory disorders, mental health and cognitive decline, bone diseases (osteoarthritis), healthy aging and quality of life among more others [16]. It has been concluded from studying the mechanism responsible for lowering these risks that food combinations, food nutrients, presence of non-nutritive substances, lifestyles habits and the cooking techniques all together make the Mediterranean dietary pattern into a tool that can not only prevent but can also be used as a way of treatment for these medical ailments [17–20]. As part of the essential dietary fat, consumption of extra virgin olive oil is the main feature of Mediterranean diet. Apart from the healthy proportion of the amount and type of fatty acids present in olive oil [21], it is consumed along with a large amount of vegetables as a salad dressing for cooked or raw legumes and vegetables (**Figure 2**). Even though the inhabitants of the Mediterranean countries lack knowledge regarding the benefits of olive oil for human health, they have believed that the long lives of their populations are attributed to the usage of olive oil in their diets.

3. Impact of olive oil consumption on human health

It has been revealed from the studies by The Global Burden of Disease that the rate of incidence of chronic illnesses such as cardiovascular diseases including coronary artery disease and hypertension have been persistently increasing [22]. It has been revealed that this increase in the rate of these diseases is most likely related to the adoption of unfavorable behaviors and unhealthy dietary patterns [23]. Therefore, the protective qualities of certain dietary patterns such as Mediterranean diet especially its certain components such as olive oil has been studied and evaluated for their beneficial effects on human mortality and morbidity [6]. Olive oil is noted to have anti-bacterial characteristics [11], involved in improving the endothelial function in young females [24], and it has been hypothesized to have epigenetic effects interplay offering protection from cancers [25] due to the presence of healthful monounsaturated fats. The presence of antioxidants contributes to the anti-inflammatory properties of the olive oil. Olive oil has high quantities of antioxidants and offers numerous benefits for cardiovascular health [26], such as protection of LDL from oxidation and lowering of the high blood pressure as well as offers protection from diabetes mellitus [27].

4. Mechanism of actions of virgin olive oil as part of Mediterranean diet to prevent cardiovascular disorders

Atherosclerosis development is becoming one of the major risk factors in people having hypertension, diabetes and hypercholesterolemia [28–30]. During the pathogenesis of hypertension, inflammation causing factors imposes oxidative stress by activating the mononuclear leucocyte by enhancing the production of wide range of cytokines which directly exert oxidative stress on the blood vessels and might lead to the development of atherosclerosis. It was reported that hypertension may develop due to variations in the arterial vasodilation/vasoconstriction particularly due to the modifications of the synthesis of molecules like NO, a potent relaxing factor, and ET1 a potent vasoconstrictor released by endothelium. It was reported in non-smoking women that when Mediterranean diet containing EVOO was given to these women for 1 year, this caused a decrease in hypertension with higher level of serum NO along with changes in gene expression including eNOS upregulation that played important role in the regulation of blood pressure, dilate the blood vessels with a number of vaso-protective and anti-atherosclerotic roles; and caveolin 2 down-regulation was also reported. Caveolin 2 can trap the eNOS and ultimately inhibit the enzyme [31]. Moreover, in patients at high risk of cardiovascular disorders when given EVOO as part of MeDiet, a significantly decrease in the systolic and diastolic blood pressure were reported [31–33] and same effects were also reported in metabolic syndrome patients for giving the MeDiet with EVOO for parallel 2 years [34].

Storniolo et al. [31] explored very interesting findings when used MeDiet containing EVOO both in moderate hypertensive and healthy volunteers and found that EVOO did not decrease the blood pressure in healthy individuals although a drop in blood pressure in moderate hypertensive patients was reported. Approaches to explore the nutritional values of MeDiet supplemented with EVOO is one of the most interesting side as EVOO have potential to reduce the development of chronic disorders like type 2 diabetes, metabolic syndrome an obesity by regulating the metabolism of glucose and lipids in our body. It was reported that hyperglycaemic and hypercholesterolemic conditions are leading cause of the development of cardiovascular disorders or aggravating the atherosclerotic process. Violi et al. [35] explored that administration of MeDiet containing the EVOO significantly regulate the postprandial metabolic profile including the decreased blood glucose, LDL-C and ox-LDL levels and an increased level of insulin was also reported in healthy volunteers. Moreover, the impact of EVOO (10 g) in MeDiet has the potential to normalize the blood glucose level in post-prandial glycaemic patients with impaired fasting glucose. The mechanisms accounting for the positive effect of EVOO are related to increasing up-regulation, as EVOO reduces dipeptidyl peptidase-4 activity with consequent increase in glucagon-like peptide-1 concentration which regulates postprandial glycaemia by up-regulating insulin secretion [36].

High level of lipids contents is one of the second one major risk factor for the development of cardiac disease but MeDiet supplemented with EVOO have significant impact to reduce the total cholesterol, LDL-cholesterol (LDL-C) and non-HDL-cholesterol (non-HDL-C) levels to prevent the cardiovascular risk [36]. Hernáez et al. [37] reported on administering the MeDiet particularly containing EVOO for 1 year to patients at high cardiovascular risk and investigated an improvement in the HDL functions including the cholesterol efflux capacity, modulation in the metabolism of cholesterol, improvement in antioxidant as well as anti-inflammatory properties, and increased vasodilatory properties. Moreover, it was also found by Covas et al. [12] that taking phenolic olive oils increases the HDL-C, decreased the total cholesterol/HDL-C ratio, decreased LDL-C/HDL-C ratio as well as decrease in triglycerides (TG) level. A study carried out in patients having metabolic syndrome reported that consumption of MeDiet supplemented with EVOO for 2 years significantly reduced the blood glucose level, increase insulin, decrease total cholesterol as well as TG and improved the HDL-C level as compared to control group. Such findings explored that Mediterranean-style diet containing EVOO is an emerging therapeutic strategy to modulate cardiovascular disease risk in metabolic syndrome [34]. During an experimental study it was also found that EVOO along with corn oil have role in improving the total cholesterol, LDL-C, VLDL-C, Apolipoprotein-B, Apolipoprotein-A and LDL particle concentrations and this improvement was observed in more significant individuals given corn oil compared with EVOO intake. Moreover, the findings of some investigations are summarized in **Table 1** and the targets of Me Diet supplemented with EVOO to prevent the cardiovascular diseases risk factors are represented in **Figure 3**.

Mediterranean diet containing extra virgin olive oil and quantity used	Number of volunteers participated	Plan of Study Design with time period used	Targets to manage Cardiovascular disorders	References
8 gram/Day	Having metabolic disorders (n = 180)	All given in parallel upto 2 years	**By Decreasing:** BMI, waist circumference, Body weight, Blood pressure, Endothelial function score, Blood glucose and insulin, hs-CRP, IL-6, IL-7, Total cholesterol, triglycerides, IL-18, **By Increasing:** HDL-C,	[34]
50 mL/Day	At high risk of cardiovascular (n = 164)	All given in parallel upto 1 year	**By Decreasing:** CRP, IL-6, [sICAM-1], [sP-selectin], [sVCAM-1], IL-18/IL-10	[38]
50 mL/Day	At high risk of cardiovascular (n = 165)	All given in parallel upto 3–5 year	**By Decreasing:** CRP, Interleukin-6, TNF-α, Total cholesterol, LDL-C, Triglyceride, Blood pressure **By Increasing:** HDL-C	[39]
1 L/ 07 Days	At high risk of cardiovascular (n = 69)	All given in parallel upto 5 years	**By Decreasing:** IL-6, IL-7, IL-8, IL-18, IL-1 β, IL-5, IL-12p70, IFN-γ, TNF-α, GCSF, GMCSF, ENA78, MCP-1, MIP-1β,	[40]
14.8 mL/Day	Given to healthy people (n = 137)	All given in parallel upto 6 months	**By Decreasing:** Systolic blood pressure **By Increasing:** FMD	[32]
1 L/07 Days	At high risk of cardiovascular (n = 231)	All given in parallel upto 1 year	**By Decreasing:** Plasma tryptophan	[41]

Table 1.
Scientific based investigations carried out to investigate the Mediterranean diet containing extra virgin olive oil on preventing the cardiovascular disorders.

DOI: http://dx.doi.org/10.5772/intechopen.97146

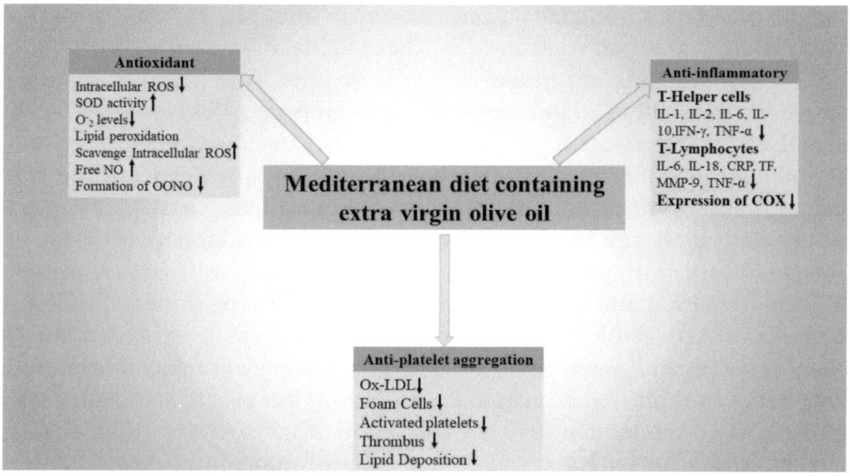

Figure 3.
Major targets of Mediterranean diet containing extra virgin olive oil to prevent the development of cardiovascular diseases.

5. Olive oil for cardiovascular diseases in secondary prevention

If the effect of the Mediterranean diet and olive oil in primary prevention appears almost obvious, it is surprising to be in secondary prevention in cardiovascular diseases. The Lyon Diet Heart Study provided information about outcomes after acute myocardial infarction. This randomized, secondary prevention trial in patients analyzed the incidence of myocardial infarction and cardiovascular deaths. Mediterranean diet was able to significantly reduce the occurrence of all-cause and cardiovascular death [42]. PREDIMED study pointed out the reduction in CD40 expression on monocytes surface, and a further decrease in C-reactive protein and IL-6 in Mediterranean diet group (plus olive oil and plus nuts) with a significant reduction in the expression of adhesion molecules. These molecular changes suggest a protection of atherosclerotic plaques from instability resulting in a reduction in cardiovascular events after a heart attack [38].

However, it is surprising how the effect of olive oil is also protective against the incidence of arrhythmias as demonstrated in the laboratory by Bukhari group [43] and as demonstrated by the sub analysis of the PREDIMED trial suggests that extra virgin olive oil in the context of a Mediterranean dietary pattern may reduce the risk of atrial fibrillation [43].

However, it will be exciting to study the PREDIMAR project, a randomized, single-blind trial testing the effect of a Mediterranean diet enriched with extra virgin olive oil to reduce tachyarrhythmia relapses after atrial fibrillation ablation. If the study hypothesis will be confirmed, the utility of the Mediterranean diet enriched with extra virgin olive oil in slowing the progression of AF will be proven and it could actually become a true healing weapon, on a par with the most recent interventional cardiological technological innovations [44].

6. Conclusion

It has been concluded from studying the mechanism responsible for lowering the cardiovascular risks that food combinations, food nutrients, presence of non-nutritive substances, lifestyles habits and the cooking techniques all together make the Mediterranean dietary pattern into a tool that can not only prevent but can also be used as a way of treatment for these medical ailments. As part of the essential dietary fat, consumption of extra virgin olive oil is the main feature of Mediterranean diet. Olive oil possess anti-bacterial properties, involved in improving the endothelial function in young females. It has been hypothesized that the olive oil has epigenetic effects offering protection against cancers due to the presence of monounsaturated fats. The presence of antioxidants contributes to the anti-inflammatory properties of the olive oil. Olive oil has high quantities of antioxidants and offers numerous benefits for cardiovascular health, such as protection of LDL from oxidation and lowering of high blood pressure. Various studies have shown that the intake of olive oil offers cardioprotective properties contributed by the high poly phenolic content as well as the presence of antioxidants and its curative effect is not limited to primary prevention, but also to secondary prevention in unexpected fields such as in the treatment of cardiac arrhythmias. Olive oil has many biological activities due to the presence of many important secondary metabolites such as antioxidant and other useful bioactive components that suggests the future use of olive oil in pharmaceutical preparations in pharmaceutical industries.

Author details

Muhammad Akram[1*], Rumaisa Ansari[1], Naheed Akhter[2], Olutosin Ademola Otekunrin[3], Sadia Zafar[4], Muhammad Ishaque[1], Naveed Munir[5], Luigi Sciarra[6], Giulia My[7], E.C. Gianvito Matarrese[8,9], Zefferino Palamà[7] and Muhammad Riaz[10]

1 Department of Eastern Medicine, Government College University Faisalabad, Pakistan

2 College of Allied Health Professional, Government College University Faisalabad, Pakistan

3 Department of Agricultural Economics and Farm Management, Federal University of Agriculture, Abeokuta (FUNAAB), Nigeria

4 Department of Botany, Division of Science and Technology, University of Education Lahore, Punjab, Pakistan

5 Department of Biochemistry, Government College University Faisalabad, Pakistan

6 Cardiology and Electrophysiology Unit, Policlinico "Casilino", Rome, Italy

7 Cardiology and Electrophysiology Unit, Villa Verde Hospital, Taranto, Italy

8 University of Bilkent, Ankara, Turkey

9 EVO Restaurant, Alberobello, Italy

10 Department of Allied Health Sciences, University of Sargodha, Pakistan

*Address all correspondence to: makram_0451@hotmail.com

References

[1] Nomikos NN, Nomikos GN and Kores DS. The use of deep friction massage with olive oil as a means of prevention and treatment of sports injuries in ancient times. *Archives of medical science: AMS*. 2010; 6: 642.

[2] Rossi R. The Eu Olive and Olive Oil Sector Main Features, Challenges and Prospects. *European Parliamentary Research Service; European Parliament: Brussels, Belgium*. 2017: 12.

[3] Vossen P. Olive oil: history, production, and characteristics of the world's classic oils. *HortScience*. 2007; 42: 1093-100.

[4] La Lastra C, Barranco M, Motilva V and Herrerias J. Mediterrranean diet and health biological importance of olive oil. *Current pharmaceutical design*. 2001; 7: 933-50.

[5] Grosso G, Marventano S, Yang J, et al. A comprehensive meta-analysis on evidence of Mediterranean diet and cardiovascular disease: are individual components equal? *Critical reviews in food science and nutrition*. 2017; 57: 3218-32.

[6] Buckland G and Gonzalez CA. The role of olive oil in disease prevention: a focus on the recent epidemiological evidence from cohort studies and dietary intervention trials. *British Journal of Nutrition*. 2015; 113: S94-S101.

[7] Gorzynik-Debicka M, Przychodzen P, Cappello F, et al. Potential health benefits of olive oil and plant polyphenols. *International journal of molecular sciences*. 2018; 19: 686.

[8] Bianchi G. Lipids and phenols in table olives. *European Journal of Lipid Science and Technology*. 2003; 105: 229-42.

[9] Tripoli E, Giammanco M, Tabacchi G, Di Majo D, Giammanco S and La Guardia M. The phenolic compounds of olive oil: structure, biological activity and beneficial effects on human health. *Nutrition research reviews*. 2005; 18: 98-112.

[10] Bogani P, Galli C, Villa M and Visioli F. Postprandial anti-inflammatory and antioxidant effects of extra virgin olive oil. *Atherosclerosis*. 2007; 190: 181-6.

[11] Servili M, Esposto S, Fabiani R, et al. Phenolic compounds in olive oil: antioxidant, health and organoleptic activities according to their chemical structure. *Inflammopharmacology*. 2009; 17: 76-84.

[12] Covas M-I, de la Torre R and Fitó M. Virgin olive oil: a key food for cardiovascular risk protection. *British Journal of Nutrition*. 2015; 113: S19-S28.

[13] Papadopoulos G and Boskou D. Antioxidant effect of natural phenols on olive oil. *Journal of the American Oil Chemists' Society*. 1991; 68: 669-71.

[14] Bondioli P, Mariani C, Lanzani A, Fedeli E, Mossa A and Muller A. Lampante olive oil refining with supercritical carbon dioxide. *Journal of the American Oil Chemists' Society*. 1992; 69: 477-80.

[15] Keys A, Mienotti A, Karvonen MJ, et al. The diet and 15-year death rate in the seven countries study. *American journal of epidemiology*. 1986; 124: 903-15.

[16] Dinu M, Pagliai G, Casini A and Sofi F. Mediterranean diet and multiple health outcomes: an umbrella review of meta-analyses of observational studies and randomised trials. *European journal of clinical nutrition*. 2018; 72: 30-43.

[17] Corella D, Coltell O, Macian F and Ordovás JM. Advances in understanding the molecular basis of

the mediterranean diet effect. *Annual review of food science and technology*. 2018; 9: 227-49.

[18] Radd-Vagenas S, Kouris-Blazos A, Singh MF and Flood VM. Evolution of Mediterranean diets and cuisine: concepts and definitions. *Asia Pacific journal of clinical nutrition*. 2017.

[19] Vasto S, Buscemi S, Barera A, Di Carlo M, Accardi G and Caruso C. Mediterranean diet and healthy ageing: a Sicilian perspective. *Gerontology*. 2014; 60: 508-18.

[20] Ortega R. Importance of functional foods in the Mediterranean diet. *Public health nutrition*. 2006; 9: 1136-40.

[21] Garcia-Martinez O, Ruiz C, Gutierrez-Ibanez A, Illescas-Montes R and Melguizo-Rodriguez L. Benefits of olive oil phenolic compounds in disease prevention. *Endocrine, Metabolic & Immune Disorders-Drug Targets (Formerly Current Drug Targets-Immune, Endocrine & Metabolic Disorders)*. 2018; 18: 333-40.

[22] Vos T, Barber RM, Bell B, et al. Global, regional, and national incidence, prevalence, and years lived with disability for 301 acute and chronic diseases and injuries in 188 countries, 1990-2013: a systematic analysis for the Global Burden of Disease Study 2013. *The Lancet*. 2015; 386: 743-800.

[23] Wagner K-H and Brath H. A global view on the development of non communicable diseases. *Preventive medicine*. 2012; 54: S38-S41.

[24] Moreno-Luna R, Muñoz-Hernandez R, Miranda ML, et al. Olive oil polyphenols decrease blood pressure and improve endothelial function in young women with mild hypertension. *American journal of hypertension*. 2012; 25: 1299-304.

[25] Menendez JA and Lupu R. Mediterranean dietary traditions for the molecular treatment of human cancer: anti-oncogenic actions of the main olive oil's monounsaturated fatty acid oleic acid (18: 1n-9). *Current pharmaceutical biotechnology*. 2006; 7: 495-502.

[26] Schwingshackl L, Christoph M and Hoffmann G. Effects of olive oil on markers of inflammation and endothelial function—a systematic review and meta-analysis. *Nutrients*. 2015; 7: 7651-75.

[27] Guasch-Ferré M, Hruby A, Salas-Salvadó J, et al. Olive oil consumption and risk of type 2 diabetes in US women. *The American journal of clinical nutrition*. 2015; 102: 479-86.

[28] Hollander W. Role of hypertension in atherosclerosis and cardiovascular disease. *The American journal of cardiology*. 1976; 38: 786-800.

[29] Choy P. Siow YL, Mymin D, OK. *Lipids and atherosclerosis Biochem Cell Biol*. 2004; 82: 212-24.

[30] Selvin E, Coresh J, Golden SH, Boland LL, Brancati FL and Steffes MW. Glycemic control, atherosclerosis, and risk factors for cardiovascular disease in individuals with diabetes: the atherosclerosis risk in communities study. *Diabetes care*. 2005; 28: 1965-73.

[31] Storniolo CE, Casillas R, Bulló M, et al. A Mediterranean diet supplemented with extra virgin olive oil or nuts improves endothelial markers involved in blood pressure control in hypertensive women. *European journal of nutrition*. 2017; 56: 89-97.

[32] Davis CR, Hodgson JM, Woodman R, Bryan J, Wilson C and Murphy KJ. A Mediterranean diet lowers blood pressure and improves endothelial function: results from the MedLey randomized intervention trial. *The American journal of clinical nutrition*. 2017; 105: 1305-13.

[33] Medina-Remón A, Tresserra-Rimbau A, Pons A, et al. Effects of total dietary polyphenols on plasma nitric oxide and blood pressure in a high cardiovascular risk cohort. The PREDIMED randomized trial. *Nutrition, Metabolism and Cardiovascular Diseases*. 2015; 25: 60-7.

[34] Esposito K, Marfella R, Ciotola M, et al. Effect of a Mediterranean-style diet on endothelial dysfunction and markers of vascular inflammation in the metabolic syndrome: a randomized trial. *Jama*. 2004; 292: 1440-6.

[35] Violi F, Loffredo L, Pignatelli P, et al. Extra virgin olive oil use is associated with improved post-prandial blood glucose and LDL cholesterol in healthy subjects. *Nutrition & diabetes*. 2015; 5: e172-e.

[36] Reiner Ž, Catapano AL, De Backer G, et al. ESC/EAS Guidelines for the management of dyslipidaemias: the Task Force for the management of dyslipidaemias of the European Society of Cardiology (ESC) and the European Atherosclerosis Society (EAS). *European heart journal*. 2011; 32: 1769-818.

[37] Hernáez Á, Castañer O, Goday A, et al. The Mediterranean Diet decreases LDL atherogenicity in high cardiovascular risk individuals: a randomized controlled trial. *Molecular nutrition & food research*. 2017; 61: 1601015.

[38] Casas R, Sacanella E, Urpi-Sarda M, et al. The effects of the mediterranean diet on biomarkers of vascular wall inflammation and plaque vulnerability in subjects with high risk for cardiovascular disease. A randomized trial. *PloS one*. 2014; 9: e100084.

[39] Casas R, Sacanella E, Urpí-Sardà M, et al. Long-term immunomodulatory effects of a mediterranean diet in adults at high risk of cardiovascular disease in the PREvención con DIeta MEDiterránea (PREDIMED) randomized controlled trial. *The Journal of nutrition*. 2016; 146: 1684-93.

[40] Casas R, Urpi-Sardà M, Sacanella E, et al. Anti-inflamsmatory effects of the Mediterranean diet in the early and late stages of atheroma plaque development. *Mediators of inflammation*. 2017; 2017.

[41] Yu E, Ruiz-Canela M, Guasch-Ferré M, et al. Increases in plasma tryptophan are inversely associated with incident cardiovascular disease in the Prevencion con Dieta Mediterranea (PREDIMED) Study. *The Journal of nutrition*. 2017; 147: 314-22.

[42] De Lorgeril M, Salen P, Martin J-L, Monjaud I, Delaye J and Mamelle N. Mediterranean diet, traditional risk factors, and the rate of cardiovascular complications after myocardial infarction: final report of the Lyon Diet Heart Study. *Circulation*. 1999; 99: 779-85.

[43] Bukhari IA, Mohamed OY, Almotrefi AA, et al. Cardioprotective Effect of Olive Oil Against Ischemia Reperfusion-induced Cardiac Arrhythmia in Isolated Diabetic Rat Heart. *Cureus*. 2020; 12.

[44] Barrio-Lopez M, Ruiz-Canela M, Ramos P, et al. PREvention of recurrent arrhythmias with Mediterranean diet (PREDIMAR) study in patients with atrial fibrillation: Rationale, design and methods. *American heart journal*. 2020; 220: 127-36.

Chapter 6

Olive Oil Production in Albania, Chemical Characterization, and Authenticity

Dritan Topi, Gamze Guclu, Hasim Kelebek and Serkan Selli

Abstract

Olive tree is present to the Western and Southern regions of Albania, alongside Adriatic and Ionian Sea, two body waters of the Mediterranean basin. Genetic studies have revealed the existence of 22 native olive cultivars, while several introduced foreign olive cultivars are present. Two most important olive cultivars respectively, exploited in the olive oil production, and table olive, are *Kalinjot* and *Kokërrmadh Berati*. Olive fruit production ranks the country 20th in the world. Olive tree comprises an important permanent crop with considerable potential for the Albanian economy. Principal component analyses (PCA) of fatty acids in OO displays their differentiation according to the cultivar and their region. Chemometric analysis gives support to the differentiation of OO according to the olive *cv.* in terms of phenolic compounds. Secoiridoids are found in abundance, *3,4-DHPEA-EDA* and *p-HPEA-EDA* as dominant compounds, especially in *Kalinjot* olive oils. Albanian OO shows high levels of aroma compounds with (*E*)-2-hexenal as the principal aroma compound. Its concentrations reach up to 40411 μg/kg in *Kalinjot cv.,* much higher compared to *Bardhi Tirana* cv (27542.7 μg/kg). The authenticity of OOs constitutes an opportunity for domestic production and certification according to the geography or origin and present an important resource to the development of a sustainable economy.

Keywords: Olive oil, *Kalinjot*, *Bardhi Tirana*, Phenolic compounds, Albania

1. Introduction

In the last decades, the interest in olive tree (*Olea europaea* L.), an important crop for Mediterranean countries, has been extended to other regions of the world, such as Australia, North and South America, due to its valuable products; olive oil and table olive [1]. *Olea europea* L. is an evergreen plant-derived from tropical and subtropical species. It is native to the Mediterranean region, tropical and central Asia, and other parts of North Africa. Fossil remains of olive species have been found in Italy, France, and many other countries. Olive tree includes many clusters and more than 2600 cultivars, many of which may be ecotypes. Native olive *O. oleaster* and cultivated olive *O. sativa* are the main species in the Mediterranean [2].

The geography of Albania has shaped the climate characteristics, with South and Western regions typical of the Mediterranean climate while the Eastern and Northern regions with typical continental climate (**Figure 1**). Olive tree is cultivated mainly in the regions with Mediterranean climate, by penetrating the

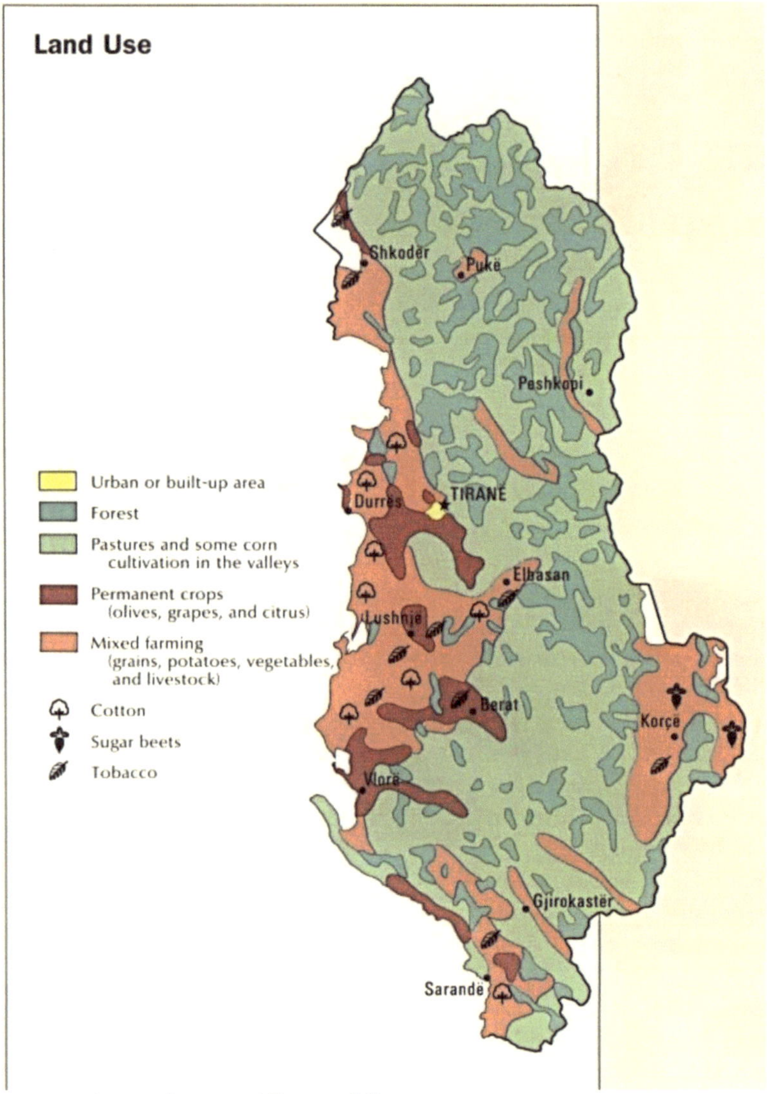

Figure 1.
Land use in Albania [3].

mainland through the river valleys, up to 560 m altitudes. Genetic studies have found 22 native olive cultivars strictly distributed into six regions - Berat, Elbasan, Kruja, Lezha, Tirana and Vlora. Native cultivars, namely: *Boç, Frëng, Kalinjot, Karre, Kushan, Kotruvs, Kokërrmadh Berati, Mixan, Kokërrmadh Elbasani, Micka, Krips, Managjel, Nisjot, Peperri, Ulli deti, Ulli i zi, Ulli bardhë Lezha, Sterbjak, Ulli i kuq, Bardhi Kruja, Bardhi Tirana (Bianco di Tirana)*, and *Ulli i bardhë Berati* [4, 5].

According to their area of plantation, they are classified as 'Principal' and 'Secondary' cultivars. *Kalinjot* is the most distinguished native olive cultivar and is distributed in over 50% of the country's plantation area. Other native cultivars important for the local economies are *Bardhi Tirana* and *Mixan*. *Kokërrmadh Berati* cultivar is the primary representative and consequently the most abundant table olive in Albania. The remaining cultivars presented in the study were classified as Secondary cultivars [5]. Olive cultivation is closely linked with local comunities for a long period. Olive cultivar names are conected with regions, as well as the olive fruit name, in albanian (ulli).

1.1 Olive tree cultivation and production

The olive tree dominates the country's permanent crops with 10.28 million in a total of 13.82 million fruit trees, or 74.4%. The main regions where the olive tree is cultivated are Vlora, Fieri, Berat, and Elbasan (**Table 1**) [6, 7].

The entire agriculture sector and especially permanent crops suffered in the last in the alternation of the 20th century with the 21st the consequences of economic transition (**Figure 2**). The arable land reform distributed the entire area to the small farms. This was a big reverse step toward the permanent crop's cultivation, which suffered the most, with negative consequences such decrease in tree

No	District	Fruit trees	Olives	Citrus	Vineyards (Ha)	Olive share/district
1	**Berat**	1,132	1,949	132	1,155	18.9
2	Dibër	1,521	—	—	2g00	0.0
3	**Durrës**	810	442	102	767	4.3
4	**Elbasan**	1,401	1,644	72	1,368	16.0
5	**Fier**	1,579	2,617	381	2,114	25.4
6	**Gjirokastër**	326	303	2	798	2.9
7	Korçë	3,228	—	—	1,117	0.0
8	Kukës	869	—	—	88	0.0
9	**Lezhë**	460	239	46	406	2.3
10	**Shkodër**	888	425	46	729	4.1
11	**Tiranë**	927	895	99	845	8.7
12	**Vlorë**	680	1,773	642	1,256	17.2
	Country	**13,822**	**10,288**	**1,521**	**10,842**	

Source: [6].

Table 1.
Permanent crops data in 2019 (× 1000).

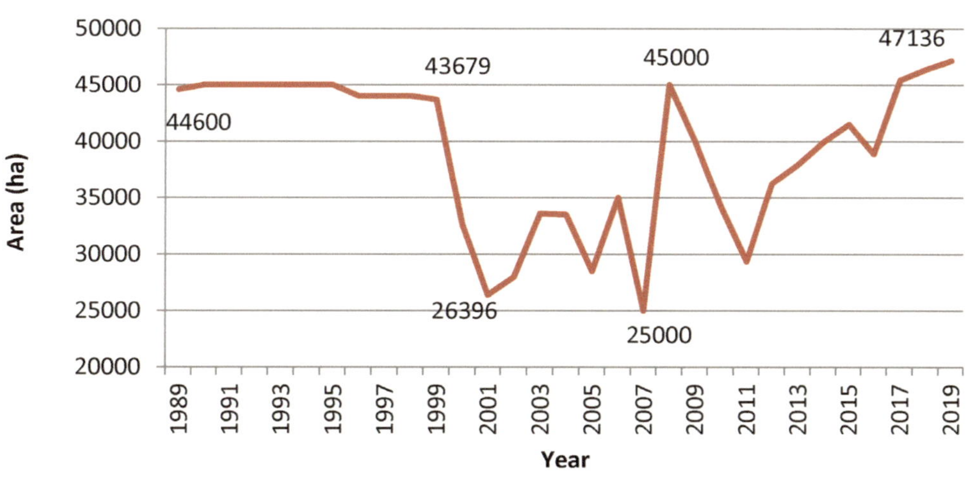

Figure 2.
Area of olive trees during 1989–2019 (in ha) [6].

numbers, and productivity. Only around 2010 because of the central government policy, focusing the priority on agriculture, and especially to crops that countries already have high productivity, the interest in olive culture highly increased. A steady increase was observed in the number of trees and the area planted. Currently, over 47000 ha are planted with olive tree, or 7.6% of the arable land [6, 8].

FAO statistics on olive fruit production ranks Albania the 20th in the world, with 98313 tons. The world production during 2019 harvesting year reached 19.46 million tons, with Albania's contribution of 0.51% [8]. A total of 8.23 out of 10.28 million olive trees are under production [6]. The olive oil production was calculated to as 20,038 tons, contribution up to 95% of domestic demands. Data on olive oil consumption increased to five liters per capita, being the highest level among the non-European Union member countries [9]. With an average production varying to 40,000 tons olive fruit, a sharp increase was evident in 2008 as well as in the following years (**Figure 3**) [10]. A key role to this positive trend was the application of support schemes subsidizing local farmers. Currently the annual production fluctuates in 100000 tons of olives. Intensive production belongs to three regions: Berat (38,000 tons), Fier (33,000 tons) and Vlora (17,000 tons), with figures belonging to 2019 [6].

1.2 Table olive production

The olive fruits are exploited in 79 percent for olive oil production, while the remaining goes to table olives production. Country holds the global record for table olive consumption per capita. According to International Olive Council, table olive consumption was 10.8 kilograms per capita [9]. *Kokërrmadh Berati* cv is the most distinguished native cultivar used as table olive. It is mainly present in the Berati region. Other olive cultivars used as table olive are not industrially processed but used only locally by families. Another important olive cultivar, *Kalinjot* is used both to extract olive oil, as well as table olive. Their nutritional profiles show small differences regarding to the main fatty acids (**Table 2**). Their n/6/n-3 ratio is around 10, showing very good profiles of unsaturated FA, compared to other distinguished olive cultivars worldwide [11]. Both are classified as High-Content Oleic Acid cultivars.

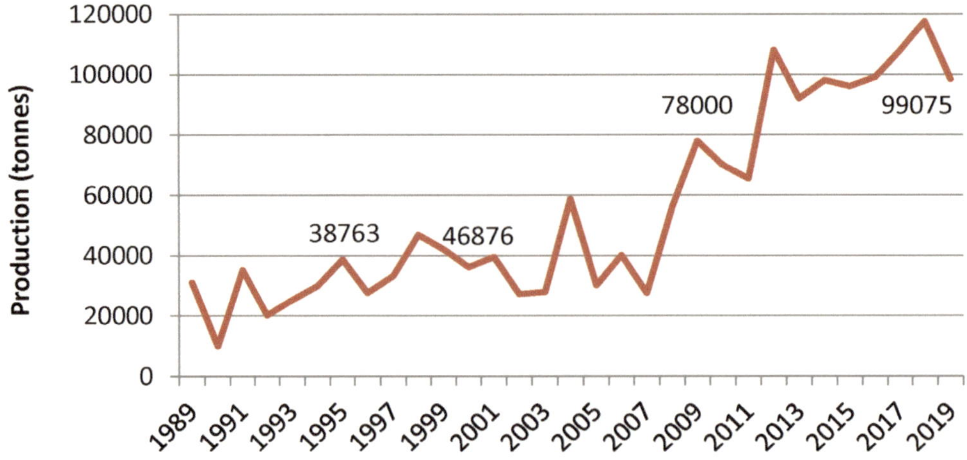

Figure 3.
Annual production for olive fruits during 1989–2019, in Albania [6].

Formula	Kalinjot	*Kokërrmadh Berati*
16:0	10.92	10.41
16:1(n-9)	0.09	0.13
16:1(n-7)	0.48	0.61
17:0	0.04	ND
17:1 (n-7)	0.14	ND
18:0	2.31	2.10
18:1(n-9)cis	75.11	76.26
18:1(n-7)	1.88	2.20
18:2 (n-6)cis	7.56	6.92
20:0	0.36	0.40
18:3 (n-3)	0.72	0.67
20:1 (n-9)	0.31	0.33
22:0	0.08	ND
n-6/n-3	10.47	10.31
Σ-SFA	13.36	12.92
Σ-MUFA	78.01	79.53
Σ-PUFA	8.29	7.59

Table 2.
Fatty acid profiles and nutritional interest of two main table olive cultivars (%).

1.3 Olive oil extraction industry in Albania

OO extraction is organized in small extraction mills, with only 1/3 of their extraction capacity exploited. OO production is reached mainly in Southern and Central parts of the country, comprising Vlora, Fieri, Berati, Elbasani and Tirana regions. These regions, in total they comprise more than 80% of the olive trees in production [12, 13]. The extraction process produces high amounts of Olive Husk (OH) and Olive Mill Wastewaters (OMWW). Their disposal in the environment is a critical issue to the Mediterranean countries. It has been shown that the OMWW disposal into surface waters influence negatively in their biodiversity due to high organic load and toxic substances. The OMWW composition varies qualitatively and quantitatively according to the olive variety, climate condition, cultivation practices, the olive storage time and olive extraction process. Composed of 83–92% water, 4–16% organic matter, and 1–2% minerals, constitute a potential for exploitation as irrigation source and fertilizer in arable lands [13, 14]. Solid residues show high interest if they are used in soil enrichment with potassium (K) and other minerals. Up to 254.85 mg/kg of K, 20 mg/kg magnesium on dry weight basis is found [14, 15].

Regions where the extraction lines are stationed face the deteriorating situation in the environmental conditions. OMWW discharge to surface waters, solid waste disposal produces bad odors on a large perimeter, disturbing the community. The OMWW amounts produced is 125–137 $\times$ 10^6 kg, while the OH approximately to 60 $\times$ 10^6 kg with a fluctuation in a yearly basis [15]. Traditionally, olive husk is used as feedstuff to animals or burned to farmers' houses. Recently, showing the high interest for exploitation of OH by-products, an investment in processing plant is operating at capacity 2.5 ton/hour for production of pellets.

2. Olive oils characterization and authenticity

2.1 *Kalinjot cv.* Olive oils

Unique in its nutritional and sensory characteristics [16], VOO plays a vital role as the primary source of fats in the traditional Mediterranean diet [1, 17].

Kalinjot, the most important olive cultivar, gives the main contribution to the national level on the OO production. It covers 70% of the plantations' structure, to Vlora and Mallakastra regions, and relatively resilient to drought and cold weather (**Figure 4**). Fruit and stone weight, respectively to approximate values of 3.6 g and 0.5 g, with the extractability rate that varies up to 28% w/w. Fatty acid composition for olive oils obtained from different locations from Vlora and Mallakastra regions, in different harvesting years the last decade (**Table 3**) [18–22].

The main fatty acid, Oleic acid, interval is 68.03–76.83%; with linoleic acid interval 7.85–14.22% and palmitic acid interval 8.54–13.62%. The Linolenic acid content range is 0.63–0.89% lying under the value established by EU legislation to olive oil (**Table 3**).

As expected, no statistically significant differences were observed between fatty acids of Kalinjot olive oils, and therefore the data presented refer to all samples analyzed. The distribution of fatty acid composition of the oil samples studied is shown in **Table 1** and covers the normal range expected for olive oil. The Kalinjot OO has a high percentage of oleic acid, with an average value of 74.25 and an interquartile range of 1.899 (the difference between samples percentiles), and a low percentage of linoleic acid, with an average value of 9.788 and an interquartile range of 1.471.

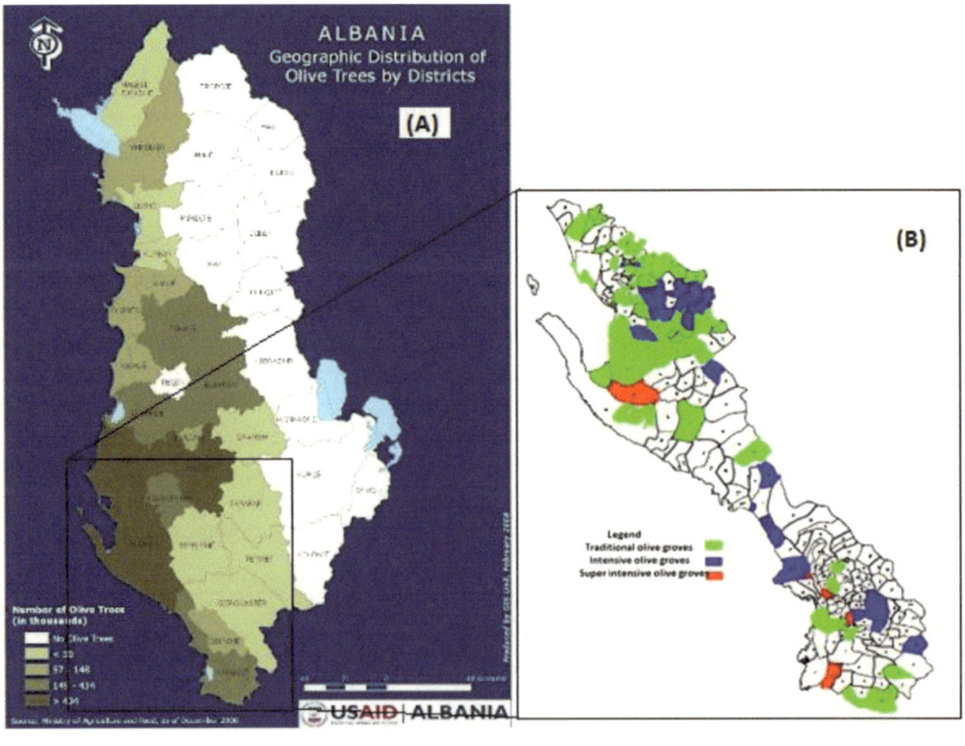

Figure 4.
(A) Olive tree cultivation regions in Albania; (B) Vlora region, classifyed as: traditional, intensive and superintensive groves. Olive tree cultivation regions in Albania, Vlora region [12].

	Mean	SDev	Max	Min
C14:0	**0.00**	0.00	0.00	0.00
C16:0	**10.43**	1.45	13.62	8.54
C16:1	**0.74**	0.53	2.91	0.42
C17:0	**0.11**	0.04	0.16	0.00
C17:1	**0.18**	0.04	0.23	0.07
C18:0	**2.69**	0.29	3.39	2.04
C18:1n9tr	**0.00**	0.00	0.00	0.00
C18:1n9c	**72.44**	2.34	76.83	68.03
C18:1n6c	**1.81**	0.71	3.16	0.00
C18:2n6c	**9.79**	1.47	14.22	7.85
C20:0	**0.47**	0.03	0.53	0.40
C18:3n3	**0.77**	0.06	0.89	0.63
C20:1	**0.38**	0.03	0.43	0.30
C21:0	**0.00**	0.00	0.00	0.00
C22:0	**0.13**	0.01	0.18	0.12
C24:0	**0.06**	0.03	0.09	0.00

Table 3.
Fatty acid mean, maximum and minimum values to Kalinjot OO.

Olive cultivar	Mean
Mixan	139.24 ± 6.56
Frëng	42.78 ± 7.04
Bardhi Kruja	322.05 ± 5.61
Kalinjot	285.16 ± 3.29
Bardhi Tirana	445.03 ± 16.83
Karren	89.74 ± 0.47
Nisjot	203.07 ± 7.51
Kotruvs	226.97 ± 1.40
Kokërrmadh	125.60 ± 6.09

Table 4.
Total phenolic content in OO from native cv (mg gallic acid/kg).

Principal component analysis given in **Figure 5** displayed the distribution of Kalinjot OO samples from different location within regions of Vlora and Mallakastra. As it can be seen in the chart, according to the PCA biplot, the samples were well categorized. The samples coded with numbers 1, 2, 6, 7, 12, 13, 16, 19, 21, 22, 23 and 24 were positioned on the positive F1 axis while the remaining were on the negative F1 axis. In addition to those, samples coded as 29 and 30 were positioned on the negative F1 axis and clearly separated from other OO samples. OO sample (29) belong to Nisjot olive cv., another olive cv of the Vlora and Mallakastra region, while the OO sample (30) belonged to the same olive cultivar, Kalinjot, but from a different geographical region, Central Albania.

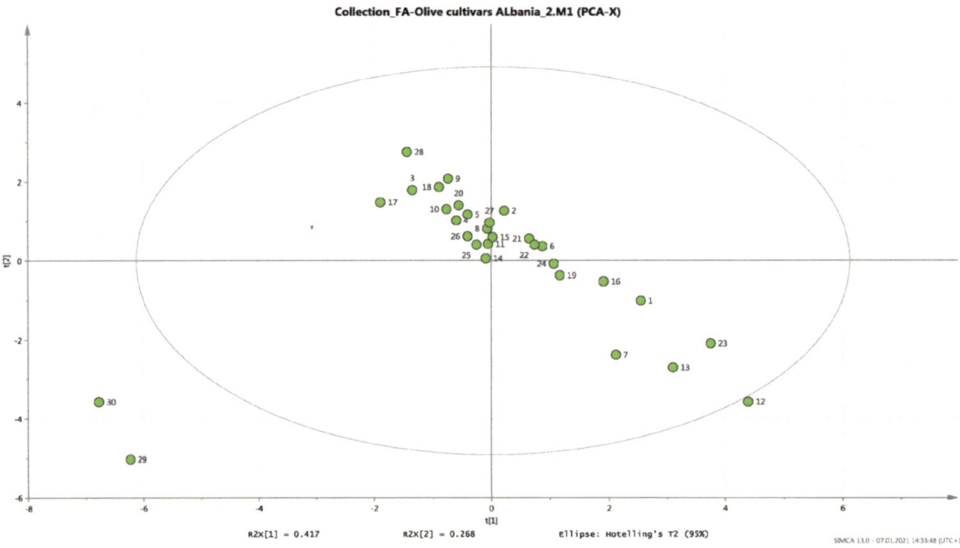

Figure 5.
Sample distribution for Kalinjot OO samples from different locations within regions of Vlora and Mallakastra.

2.2 Fatty acid composition

Fatty acid profiles from twenty-six different olive cv are presented in **Table 6**. Oleic (C18:1), palmitic (16:0), linoleic (C18:2), and stearic (C18:0) acids, are primary FA in decreasing order. Results revealed that the fatty acid content falls in the average percentage intervals described by FAO and IOOC [23]. These olive varieties presented a significant variation in oleic acid (OA). They are clustered into two groups containing Oleic acid, low-OA cultivars: *Krips Kruja* (66.24%) and *Peperri* (66.27%); and high-OA cultivars: *Kalinjot*, *Bardhi Tirana*, *Mixan*, *Frëng*, *Bardhi Kruja*, *Managjel*, *Ulli deti* (from 70.87 to 76.58%) [24]. Aparacio and Luna [25] have suggested a correlation between the cultivars' chemical composition and the pedo-climatic conditions. These variations are probably more related to genetic factors than to environmental conditions.

Palmitic acid is found at concentrations five to eight-fold lower than OA, i.e., between 9.41% (*Kalinjot*) to 12.87% (*Ulli deti*), with no considerable differences among them. The content of linoleic acid (LA) varied from 5.73% (*Frëng*) to 15.19% (*Krips Kruja*), whereas the content of alfa-linolenic acid showed a small variation from 0.38% (*Ulli deti*) to 0.94% (*Sterbjak*) [11, 24].

Phenolic compounds	*Kalinjot* (Himara)	*Kalinjot* (Vlora)	Bardhi Tirana	Ulli-i-zi	Krips Kruja	Bardhi Kruja
Phenolic alcohols	61.44	84.91	72.31	72.12	34.10	15.68
Phenolic acids	0.95	0.53	1.04	1.21	6.23	12.48
Secoiridoids	135.72	162.50	146.61	170.16	61.70	22.36
Flavonoids	0.19	0.32	0.23	15.85	2.33	0.60
Phenolic aldehydes	0.24	0.08	0.27	0.31	1.64	3.07
Total phenolics	198.54	248.34	220.46	259.65	106.00	54.20

Table 5.
Total phenolic compounds of OO from Albanian olive cultivars (mg/kg).

No.	Sample name	16:0	16:1(n-9)	16:1(n-7)	17:0	17:1 (n-7)	18:0	18:1(n-9)c	18:1(n-7)	18:2(n-6)t	18:2 (n-6)c	20:0	18:3 (n-3)	20:1 (n-9)	22:0
1	Boç	13.05	0.15	0.63	0.12	0.22	2.11	67.65	2.24	0.00	12.46	0.354	0.64	0.2375	0.12
2	Kalinjot (Mkaster)	10.92	0.09	0.48	0.04	0.14	2.31	75.11	1.88	0.00	7.56	0.36	0.72	0.3059	0.08
3	Bardhi Tirana	10.88	0.07	0.35	0.13	0.19	2.83	74.61	1.53	0.00	8.00	0.43	0.58	0.2801	0.07
4	Bardhi Lezha	12.85	0.12	0.97	0.00	0.00	2.53	70.84	2.49	0.00	8.69	0.42	0.86	0.1758	0.04
5	Sterbjak	10.89	0.15	0.64	0.00	0.00	2.21	75.12	2.10	0.00	7.23	0.36	0.94	0.24	0.04
6	Ulli I kuq	12.95	0.10	0.83	0.15	0.24	2.64	66.26	2.28	0.00	13.29	0.41	0.45	0.25	0.12
7	Peperr	12.32	0.09	0.61	0.11	0.17	2.26	66.27	2.09	0.046	14.76	0.39	0.49	0.25	0.13
8	Ulli deti	12.87	0.09	1.17	0.03	0.08	2.29	70.87	2.84	0.02	8.60	0.31	0.38	0.24	0.18
9	Mixan	12.38	0.08	0.51	0.12	0.17	2.88	71.91	1.76	0.01	8.77	0.46	0.51	0.25	0.18
10	Managjel	10.15	0.09	0.34	0.14	0.22	2.72	75.22	1.48	0.05	8.29	0.38	0.50	0.23	0.19
11	Ulli i zi Tirana	11.01	0.09	0.51	0.09	0.16	2.76	70.99	2.03	0.05	10.80	0.38	0.72	0.22	0.14
12	Micka	10.80	0.02	0.61	0.08	0.17	2.30	73.47	2.27	0.03	9.10	0.35	0.54	0.21	0.05
13	Karre	11.17	0.00	0.36	0.14	0.18	3.22	73.94	1.44	0.00	8.35	0.50	0.45	0.24	0.00
14	Kamza	9.27	0.00	0.56	0.12	0.19	2.79	80.32	1.41	0.00	4.08	0.38	0.80	0.07	0.00
15	Kushan	9.12	0.10	0.41	0.05	0.15	3.17	77.15	1.73	0.00	6.95	0.42	0.52	0.22	0.00
16	Nisjot Pobrati	9.93	0.12	0.41	0.00	0.00	2.56	80.07	1.55	0.00	4.10	0.45	0.50	0.31	0.00
17	Kotruvs	12.21	0.09	0.82	0.00	0.00	1.96	71.53	2.87	0.00	9.31	0.37	0.51	0.34	0.00
18	Kokërrmadh Berati	10.41	0.13	0.61	0.00	0.00	2.10	76.16	2.20	0.00	6.92	0.40	0.67	0.33	0.12
19	Leccino	13.09	0.10	1.17	0.00	0.00	1.83	75.12	2.98	0.00	4.68	0.27	0.52	0.22	0.08
20	Krips Kruje	12.09	0.15	0.61	0.12	0.24	2.16	66.23	2.19	0.00	15.19	0.29	0.50	0.19	0.07
21	Ulli i Bardhë Kruje	11.16	0.09	0.55	0.13	0.22	2.35	72.59	1.97	0.00	9.90	0.35	0.43	0.25	0.04
22	Frëng Kruje	9.62	0.12	0.60	0.00	0.00	3.87	76.57	2.14	0.00	5.73	0.54	0.51	0.28	0.04

No.	Sample name	16:0	16:1(n-9)	16:1(n-7)	17:0	17:1 (n-7)	18:0	18:1(n-9)c	18:1(n-7)	18:2(n-6)t	18:2 (n-6)c	20:0	18:3 (n-3)	20:1 (n-9)	22:0
23	Ulli bardhë Pobrati	11.99	0.07	0.38	0.11	0.16	3.18	70.72	1.57	0.00	10.49	0.49	0.47	0.25	0.12
24	Kalinjot (Marikaj)	9.41	0.10	0.26	0.13	0.18	2.98	74.58	1.30	0.00	9.80	0.39	0.56	0.29	0.03
25	Bardhi Tirana (Priskë)	10.71	0.08	0.36	0.12	0.18	3.23	75.62	1.48	0.00	6.99	0.49	0.48	0.24	0.00
26	Ulli I kuq	15.21	0.00	0.94	0.13	0.20	2.37	64.70	2.20	0.00	12.92	0.34	0.64	0.11	0.18
27	Kalinjot (Vlorë)	9.57	0.14	0.26	0.10	0.15	3.02	73.61	1.48	0.00	10.11	0.46	0.75	0.34	0.09
28	Frantoio	14.16	0.05	1.42	0.00	0.00	2.23	71.65	2.90	0.00	6.64	0.29	0.52	0.06	0.14
30	Kalinjot (Panaja)	8.94	0.14	0.22	0.14	0.20	3.27	75.41	1.28	0.00	8.85	0.48	0.72	0.34	0.07
31	Nisjot	15.22	0.00	0.94	0.13	0.20	2.38	64.71	2.20	0.00	13.24	0.34	0.54	0.11	0.01
33	Kalinjot (Mallakastra)	10.38	0.11	0.50	0.09	0.15	2.80	74.41	1.78	0.00	8.36	0.42	0.69	0.30	0.00
	Mean	11.44	0.09	0.61	0.08	0.13	2.62	72.69	1.99	0.01	9.04	0.40	0.58	0.24	0.08
	Max	15.22	0.15	1.42	0.15	0.24	3.87	80.32	2.98	0.05	15.19	0.54	0.94	0.34	0.19
	Min	8.94	0.0	0.22	0.0	0.0	1.83	64.7	1.28	0.0	4.08	0.27	0.38	0.06	0.0

Table 6.
Fatty acid profiles for OO rom different cultivars (%).

Results showed a similarity among Albanian studied cultivars and other cultivars from Italy and Greece [25]. Comparison of olive cultivars from this region with data of cultivars from the Southern Mediterranean showed a profile with a high content of polyunsaturated fatty acids [26, 27]. Statistical analyses revealed differences among the cultivars for individual fatty acids ($p < 0.05$) (**Table 6**): only the primary fatty acids, C16:0, C18:1 (n-9), and C18:2 (n-6), presented differences among each other statistically.

Foreign olive cultivars were introduced last century during sixties, with two Italian cv. *Frantoio* and Leccino. Nowadays this olive cultivars are well adopted. Palmitic acid concentrations are higher 13,09% (*Leccino*) and 14.16% (*Frantoio*) compared to OO from native olive cv. There is a similarity in respect to OA among two foreign cv and others, while their LA concentrations were very low 4.68% (*Leccino*) and 6.64% (*Frantoio*), compared to OO from Albanian olive cultivars. The content of alfa-LA acid was 0.52% from both foreign cultivars [19].

A similarity was found in FA profiles among Albanian OO from native cultivars and OO from Italy and Greece [24]. In our study, the comparison of olive cultivars with cultivars' data from the Southern coast of the Mediterranean Sea showed a profile with a high content of polyunsaturated fatty acids [26, 27]. Statistical analyses reveal differences among the cultivars for individual fatty acids ($p < 0.05$) (**Table 6**): only the primary fatty acids, C16:0, C18:1 (n-9), and C18:2 (n-6), presented differences among them statistically.

Different authors suggest the ratio C18:1/C18:2 on evaluating VOO's oxidative stability, with a proposed minimum accepted value of 7.0. Among Albanian OO, two olive cultivars, *Frëng,* and *Nisjot* exhibited higher values, 13.37 and 19.54, respectively, while the remaining cultivars produced monovarietal olive oil with good oxidative stability. Only two olive cultivars, *Krips,* and *Peperr,* exhibited lower values, 4.4 and 4.5, respectively [11, 19].

Principal component analysis (PCA) was applied in order to determine the relations between fatty acid compositions in the OO samples. Based on the results of the PCA, two different principal components were determined, and these two components described 95% of the total variability of the experimental data. The PCA biplot of the fatty acids detected in the OO samples is shown in **Figure 6**. According to the PCA biplot, OO samples were well categorized. As seen in

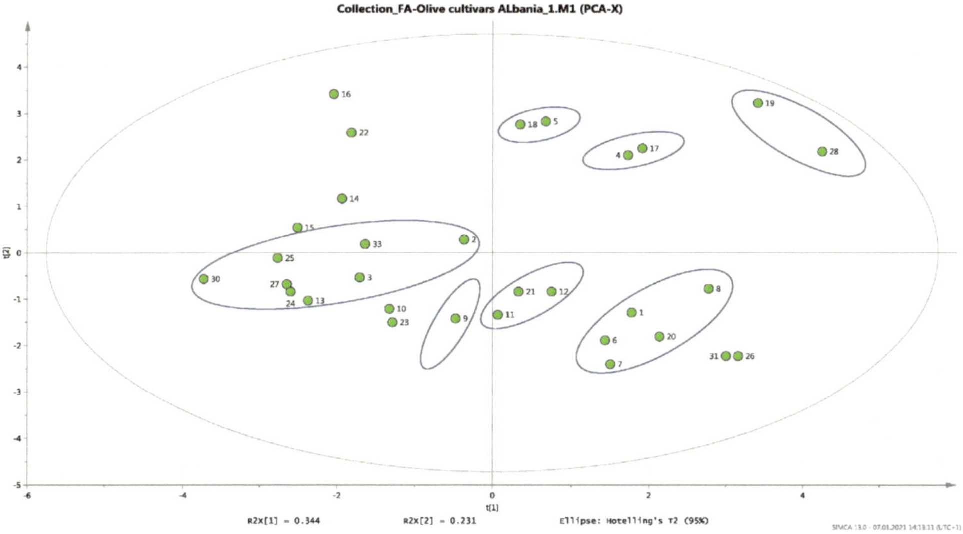

Figure 6.
PCA of FA profiles obtained from OO of different olive cultivars present in Albania.

Figure 6 and **Table 6**, the OO samples coded as 1, 4, 5, 6, 7, 8, 11, 17, 18, 19, 20, 21, 26, 28 and 31 were positioned on the positive F1 axis while the other OO samples were on the negative F1 axis.

Regarding the cultivar *Kalinjot*, OO samples have been taken from some regions of the country, because this cultivar is more widespread and more productive for olive oil. OO *Kalinjot* cv from Vlora region, no 2, and *Kalinjot* cv from Mallakastra region, no 33, fall very close, indicating the opportunity to produce OO with designated origin. OO samples no 24, 27 and 30, belonging to Kalinjot cv, fall together, telling the possibility yo produce OO from this cultivar and to authenticate them (**Table 7**). The same interesting picture belong two samples belonging to *Bardhi*

Sample nr.	Location	Harvesting season	Geographical coordinate
1	Hoshtim	2014	40° 32′ 28" North, 19° 29′ 9″ East
2	Drashovice	2014	40° 26′ 49" North, 19° 34′ 52″ East
3	Tragjas	2014	40° 19′ 32" North, 19° 30′ 36″ East
4	Trevellazer-2	2014	40° 34′ 46" North, 19° 30′ 40″ East
5	Vezhdanisht	2014	40° 29′ 11" North, 19° 36′ 37″ East
6	Kanina	2014	40° 26′ 23" North, 19° 31′ 8″ East
7	Kanina	2015	40° 26′ 23" North, 19° 31′ 8″ East
8	Panaja	2014	40° 32′ 13" North, 19° 28′ 20″ East
9	Kerkove	2014	40° 31′ 49" North, 19° 30′ 6″ East
10	Rromes	2015	40° 31′ 11" North, 19° 40′ 1″ East
11	Bestrove	2015	40° 31′ 4" North, 19° 29′ 3″ East
12	Babice	2015	40° 29′ 5" North, 19° 30′ 51″ East
13	Lubonje	2015	40° 30′ 28" North, 19° 34′ 20″ East
14	Bestrove-1	2014	40° 31′ 4" North, 19° 29′ 3″ East
15	Bestrove-2	2014	40° 31′ 4" North, 19° 29′ 3″ East
16	Himare	2014	40° 6′ 6" North, 19° 44′ 41″ East
17	Panaja	2015	40° 32′ 13" North, 19° 28′ 20″ East
18	Peshkepi	2014	40° 28′ 22" North, 19° 35′ 8″ East
19	Kanina-3	2014	40° 26′ 23" North, 19° 31′ 8″ East
20	Qeparo	2014	40° 3′ 25" North, 19° 49′ 19″ East
21	Kerkove	2014	40° 31′ 49" North, 19° 30′ 6″ East
22	Trevellazer	2014	40° 34′ 46" North, 19° 30′ 40″ East
23	Armen	2014	40° 32′ 11" North, 19° 35′ 46″ East
24	Vllahine	2014	40° 27′ 25" North, 19° 38′ 37″ East
25	Kanina-2	2014	40° 26′ 23" North, 19° 31′ 8″ East
26	Rromes	2014	40° 31′ 11" North, 19° 40′ 1″ East
27	Kanina	2015	40° 26′ 23" North, 19° 31′ 8″ East
28	Panaja	2015	40° 32′ 13" North, 19° 28′ 20″ East
29	Nisjot cv	2014	40° 33′ 32" North, 19° 40′ 53″ East
30	Marikaj-Vorë	2014	41° 22′ 20" North, 19° 37′ 56″ East

Table 7.
Sampling sites for OO Kalinjot cv. to the Vlora and Mallakastra regions..

Tirana cv, no 3 and 25, harvested from the region with the same name. OO samples respectively olive cultivars *Boç* (1), *Ulli i kuq* (6), *Peperr* (7), *Ulli deti* (8) belonging to coastal regions of Durres fall together with the olive cv from Kruja region, no 20 (*Krips Kruja*). Another interesting conclusion is similarity among OO samples belonging to olive cv from Tirana region *Ulli i zi Tirana* (11), *Micka* (12) with the olive cv *Bardhi Kruja* (21), from region with the same name. Three regions Tirana, Kruja and Durres, fall very close, Central Albania, while the olive cv show their expansion route. Through PCA box plot is found another connection among olive cv from Berat and Lezha regions. There are similarities among OO samples coded 5 (Sterbjak, Lezha region and 18, Kokerrmadh Berati, as well as OO no 4, Ulli bardhe Lezha, and 17, Kotruvs, Berati. OO no 9 belonging to Mixan cv., that belongs to Elbasan geographical region si different compared to other monocultivar OO. Very interesting is the isolation of two introduced Italian olive cv *Leccino* (19) and *Frantoio* (28).

2.3 Phenolic compounds in Albanian OO

Phenolic compounds in Olive oils usually range between 50 and 1000 ppm (mg/kg) depending on the cultivars, pedoclimatic conditions, maturity stage of the fruit, and extraction conditions [28]. OO is celebrated not only for its nutritional value but also for the content of minor compounds of pharmacologically active principles, belonging to the nutraceutical family, otherwise known as functional foods, such as polyhydroxylated phenolic and catecholic species [17] (**Figure 7**). A variety of over 230 chemical compounds, approx. 2% of the weight found in the VOO's unsaponifiable fraction, such as polyphenols, tocopherols, sterols, flavors [29–31]. Polyphenols present in VOO, are classified into two groups: lipophilic phenols (tocopherols) and hydrophilic or polar phenols [32, 33]. The hydrophilic phenols (HP) present in the VOO belong to different classes: phenolic acids, phenyl ethyl alcohols, secoiridoids, hydroxy-isochromans, flavonoids, and lignans [31]. Phenolic compounds show many health benefits, including reducing the risk factors of coronary heart disease, the prevention of several chronic diseases (for example,

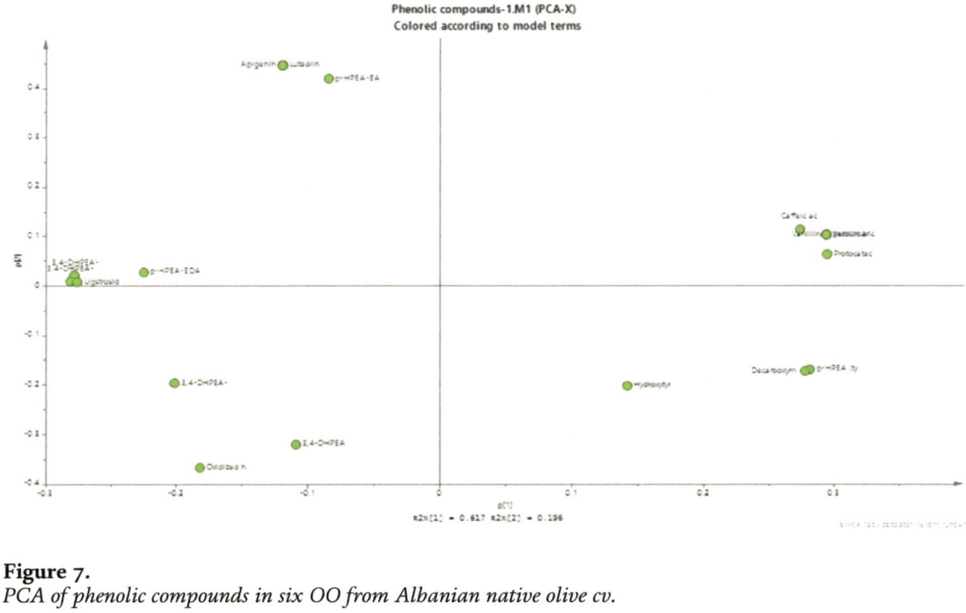

Figure 7.
PCA of phenolic compounds in six OO from Albanian native olive cv.

atherosclerosis), cancer, chronic inflammation, strokes, and other degenerative disorders [29, 34]. EFSA has concluded that polyphenols in olive are the health claims' subject. The claimed effects are "reduces oxidative stress," "antioxidant properties," "lipid metabolism," "antioxidant activity, they protect body cells and LDL from oxidative damages." An amount of 5 mg hydroxytyrosol and its derivatives (e.g., oleuropein complex and tyrosol) in olive oil should consume daily to bear the claim [35].

Results on Total Polyphenol Content (TPC) indicate that *Bardhi Tirana* variety presented the highest values, 445.03 ± 16.83 mg/kg Gallic Acid in olive oil. In contrast, *Frëng* cultivar shows the lowest levels, 42.78 ± 7.04 mg GAE/kg olive oil (**Table 4**). Results also show significant differences among cultivars, which may correlate to the cultivar, rather than agriculture practices or other factors. The comparison of TPC content in the ten studied cultivars with different cultivars, already published from Albania [18], and neighboring countries reveal that these cultivars are like cultivars from the Toscana Region (Italy) and Dalmatian Coast (Croatia) [26]. *Montedoro et al.* [36] have grouped the monovarietal olive oils according to the three groups' total phenol content. A classification for studied VOO is: "low" (50–200 mg/kg) *Ulli i Zi, Frëng, Krips, Mixan,* and *Peperri* cultivars; "medium" (200–500 mg/kg) *Bardhi Tirana, Bardhi Kruja*, *Managjel, Kalinjot,* and *Ulli deti* cultivars. The results obtained for the studied cultivars relate mainly to the cultivar differences. The results show that the studied olive oils' polyphenol content had significant differences ($p < 0.05$) among the cultivars. The actual stage of agriculture in Albania does not have a premise for irrigation of the olive plantations. Furthermore, concerning the maximum and minimum temperature values in the different regions, respectively, gives evidence that differences are not significant; hence, differences reported on the TPC content do not come due to climatic conditions. The TPC values of *Bardhi Tirana* are comparable with *Koreiniki* (Greece), *Picual* (Spain), and *Frantoio* (Italy) cultivars [25, 26].

Data on Total Polyphenol Content (TPC) indicate that OO from *Bardhi Tirana* cv present high values, 445.03 ± 16.83 mg GA/kg OO. In contrast, *Frëng* cv OO has the lowest levels, 42.78 ± 7.04 mg GA/kg (**Table 4**) [18, 19]. Total phenolic content in OO compared with OO from neighboring countries reveals that these cultivars are similar to cultivars from the Toscana Region (Italy) and Dalmatian Coast (Croatia) [26]. TPC values among studied cultivars resulted in statistically different. *Montedoro et al.* (*1992*) have grouped the monovarietal olive oils according to the three groups' total phenol content. The studied cultivars are classified as: "low" (50–200 mg/kg) *Frëng and Mixan* cultivars; "medium" (200–500 mg/kg) *Bardhi Tirana, Bardhi Kruja*, and *Kalinjot* olive cultivars. The results obtained can be related mainly to the cultivar differences. The results show that the studied olive oils' polyphenol content had significant differences ($p < 0.05$) among the cultivars. *Bardhi Tirana* samples are comparable with *Koreiniki* (Greece), *Picual* (Spain), *Frantoio* (Italy) and *Memecik*, *Ayvalik* and *Gemlik* (Turkey) cultivars [25, 26, 36].

Studying of Albanian OO from different cultivars by LC-DAD-ESI-MS/MS of has found the presence of 18 phenolic compounds. Based on their chemical structure, they were grouped as secoiridoids (6), phenolic alcohols (5), phenolic acids (4), flavonoids (2), and phenolic aldehyde (1) [37]. The highest amount of phenolic compounds was determined in *Ulli-i-Zi* cv. from Tirana region (259.65 mg/kg) followed by *Kalinjot* cv. from Vlora region (248.34 mg/kg), *Bardhi Tirana* cv. from Tirana region (220.46 mg/kg), *Kalinjot* cv. from Himara region (198.54 mg/kg), *Krips Kruja* cv. from Kruja region (106.00 mg/kg) and *Bardhi Kruja* cv. from Kruja region (54.20 mg/kg) (**Table 5**) [38].

In the study conducted by Topi et al., (2020), the results on phenolic compounds agreed with the same concentration pattern of all VOO in the literature with the highest being *3,4-DHPEA-AC* (76.03 mg/kg), followed by *3,4-DHPEA* (11.36 mg/kg), *p-HPEA* (7.01 mg/kg), oxidized hydroxytyrosol (1.53 mg/kg) and the last as hydroxytyrosol quinone (1.21 mg/kg). The highest values for 3,4-*DHPEA-AC* (76.03 mg/kg) belonged to *Kalinjot* cv. from the Vlora region, followed by *Ulli-i-Zi* cv. (65.17 mg/kg). Phenolic alcohol pattern in Albanian VOOs is consistent with published data from Spanish cv. *Picual* [39], Turkish cv. *Halhali* [40] and Croatian olive oils [41]. Hydroxytyrosol levels (3.97–11.36 mg/kg) in the *Kalinjot* cv. VOO samples, in both regions, were observed higher when compared with different Greek olive *cvs*. [40], yet exhibited a lower value compared to Croatian olive oils *Krvavica* cv. (14.9–21.9 mg/kg) [41].

Additionally, PCA was applied to generate models for the classification of OO samples in terms of total phenolic contents. It can be observed that three groups were mainly formed on the plot, of these, *Krips Kruja* and *Bardhi Kruja* cv. as a group were positioned on the positive F1 axis, while the remaining samples were located on the negative axis. The formation of this group is expected as these two *cvs*. Had the lowest values of total phenolic content. On the other hand, *Ulli-i-Zi* cv. is seen to be clearly separated from the others, expectedly as it had the highest content of phenolics.

2.4 Olive oil aroma compounds

Aroma compounds are the main criteria affecting consumer acceptance and preference and also their purchasing power remarkably. Investigation on aroma compounds in *Kalinjot* and *Bardhi Tirana* OO indicate the aroma differences between oils obtained from two main Albanian olive varieties from different regions. (*E*)-2-hexenal is found as the principal compound in both these olive *cvs.* [13]. A total of 24 aroma compounds in *Kalinjot cv*. comprising: aldehydes, alcohols, ketones, esters, terpenes, phenols, and alkenes were detected; meanwhile, in *Bardhi Tirana cv*. 17 aroma compounds including aldehydes, alcohols, esters, and phenols. The aroma compounds in Kalinjot OO varied from 36700 to 40411 μg/kg, much higher compared to Bardhi Tirana *cv*. OO (27542.7 μg/kg). Different regions and varieties play a key role in the concentration and profile of volatile compounds in the samples under study. (*E*)-2-Hexanal in *Kalinjot* (37.2–39.1%) and *Bardhi Tirana* OO (55.5%) is found lower comparing with Italian Leccino *cv*. (73%) [41], but higher compared to Croatian OO from Masnjaca cv (27.6–28.9%) [42]. Volatile compounds present in high concentrations to Kalinjot *cv.* OO were hexanal (6.6–9.8%), 3-hexenal (4.6–9.8%), 3-penten-2-ol (7.7–9.4%), (*E*)-3-hexenyl acetate (4.0–8.8%), (*Z*)-3-hexenol (2.9–4.1%), (*E*)-2-hexenol (1.4–3.1%), *β*-ocimene (n.d-4.2%) and hexanol (n.d-4.2%) [13].

Total alcohols in *Kalinjot* OO vary between 6903 and 9375 μg/kg, and in *Bardhi Tirana* OO 6874 μg/kg. The main C-5 alcohol resulted is 3-penten-2-ol in the range 2504–3442 μg/kg. This alcohol was found in higher levels compared to olive oils from Greek, Italian and Spanish olive cultivarssuch as Croatian cv. *Coratina* [43], Italian cv. *Leccino*, Spanish *cv. Cornicabra* and *Arbequina*, Greek *cv. Koroneiki* and *Adramytini* [39], and Turkish *cv. Halhali*, [44]. Total content of terpenes in Kalinjot OO varies 1895–2941 μg/kg, while the presence of *β*-ocimene may serve to distinguish OO both in olive cv and region. Terpene compounds give olive oils characteristic floral odors. (*E*)-3-Hexenyl acetate and methyl salicylate are two esters found in OO from both *Kalinjot* and *Bardh Tiranai* cultivars. Another ester found only in OO from Kalinjot is 2-butoxyethyl acetate. (*E*)-3-hexenyl acetate concentrations vary between 2705 and 3555 μg/kg in Kalinjot OO, while are found in lower levels to *Bardhi Tirana* OO (1107.7 μg/kg). The esters concentration compared with Greek cultivars is lower, however, higher when compared with Italian and Spanish *cv.* [42]. This ester compound gives olive oils pleasant and fruity odor notes. In conclusion, 2-butoxyethyl acetate may use as molecular marker to distinguish OO from these two cultivars.

3. Conclusions

Despite the modest contribution in global OO production, world ranking 20th, olive tree comprises an important permanent crop with considerable potential for the Albanian economy. OO with certified origin will increase the value of the final product. With a contribution of 0.51% to total world production, OO production reached 20,038 tons contributing almost to 95% of domestic demands. Interestingly, table olive consumption ranks the country the 1st in world. Principal component analyses (PCA) for OO samples from Kalinjot cv. displayed the differentiation with OO samples. The authenticity of OOs originating from Vlora region constitutes an opportunity for domestic production. In parallel, PCA of OO from different native and foreign olive cv. displayed differentiation regarding fatty acids.

Phenolic compounds found to be significantly different among the olive oil samples of different cultivars. *3,4-DHPEA-AC* was determined as the main phenolic compound. Secoiridoids are found in abundance, *3,4-DHPEA-EDA* and *p-HPEA-EDA* as dominant compounds, especially in *Kalinjot* OO. It was found that Albanian VOOs had lower levels of flavonoids consisting of luteolin and apigenin. Studies in OO aroma compounds from main cv indicate that Albanian OO is considered with high content OO with aroma compounds. (*E*)-2-hexenal is found as the principal aroma compound. *Kalinjot* OO concentrations vary between 36700 and 40411 μg/kg, much higher compared to *Bardhi Tirana* OO (27542.7 μg/kg).

Author details

Dritan Topi[1]*, Gamze Guclu[2], Hasim Kelebek[3] and Serkan Selli[2]

1 Department of Chemistry, Faculty of Natural Sciences, University of Tirana, Tirana, Albania

2 Department of Food Engineering, Faculty of Agriculture, Cukurova University, Adana, Turkey

3 Department of Food Engineering, Faculty of Engineering, Adana Alparslan Turkes Science and Technology University, Adana, Turkey

*Address all correspondence to: dritan.topi@unitir.edu.al

References

[1] Tous J: The influence of growing region and cultivar on olives and olive oil characteristics and their functional constituents. In: Olives and olive oil as functional foods: bioactivity, chemistry, and processing. Edited by: A. Kiritsakis, F. Shahidi. Chichester, UK, Hoboken, NJ. John Wiley & Sons, 2017.

[2] Therios I: Olives. CABI, UK; 2009. pp. 425.

[3] Map of Land Use in Albania. http://www.lib.utexas.edu/maps/atlas_east_europe/albania-landuse.jpg) University of Texas. 1990.

[4] Kafazi N: *Vecorite morfologjike dhe agrobiologjike te varieteteve te Ullirit.* Tirana: ILB, 1980.

[5] Thomaj F. Panajoti Dh. *Olive cultivars* (1 ed). Tirana: ILB; 2003. ISBN 99927 941 86.

[6] Agriculture Statistical Yearbook; 2020. Source of information: Ministry of Agriculture and Rural Development. http://www.instat.gov.al/media/3658/statistical-yearbook-2020.pdf.

[7] FAO (2018): Accessed on the web address: http://www.fao.org/faostat/en/#data/QC

[8] Wordbank; 2020. https://data.worldbank.org/indicator/AG.LND.ARBL.HA?locations=AL. (Access on 19/01/2021).

[9] OLIVE COUNCIL; 2020. https://www.internationaloliveoil.org/wp-content/uploads/2020/12/OT-W901-23-11-2020-C.pdf

[10] MACP (Ministry of Agriculture and Consumer Protection). Study on the actual olive tree cultivation and its perspectives in Albania. 2009. (in Albanian).

[11] Topi D, Thomaj F, Shabani M, Carvalho AP, Gomes AM: The analytical and nutritional study of the Albanian autochthonous table olive *Kokërmadh Berati*. *Proceedings of 6th Central European Congress on Food-CEFood Congress*. 2012a. Novi sad, Serbia.

[12] MAFCP; 2009: Study on actual Olive cultivation and its developing perspectives. Ed: Ministry of Agriculture, Food and Consumer Protection. Tirana. Albania.

[13] Topi D, Amanpour A, Kelebek H, and Selli S: Screening of aroma profiles in Albanian cvs. Kalinjot and Bardhi Tirana olive oils using purge and trap extraction technique. *RISG Rivista Italiana Sostanze Grasse*; 2019;96(2):101-108.

[14] Dermeche S, Nadour M, Larroche C, Moulti-Mati F, Michaud P: Olive mill wastes: Biochemical characterization and valorization strategies. *Process Biochemistry*; 2013; 48:1532-1552.

[15] Topi D, Beqiraj I, Seiti B, Halimi E: Environmental impact from olive mills waste disposal, chemical analysis of solid wastes and wastewaters. *Journal of Hygienic and Engineering Design*. 2014; 7: 44-48.

[16] Bendini A, Cerretani L, Carrasco-Pancorbo A, Gómez-Caravaca A. M, Segura-Carretero A, Fernández-Gutiérrez A: Phenolic Molecules in Virgin Olive Oils: A Survey of Their Sensory Properties, Health Effects, Antioxidant Activity, and Analytical Methods. An Overview of the Last Decade. *Molecules*; 2007; 12:1679-1719.

[17] Sindona G: A marker of Quality of Olive oils: The Expression of Oleuropein. In: Olives and Olive oil in Health and Disease Prevention. Editors V. R. Preedy and R. R. Watson. Elsevier. London, UK. 2010. pp. 95-100.

[18] Topi D, Thomaj F, Carvalho AP. Gomes A: Analytical Evaluation of Six

Monovarietal Virgin Olive Oils from Kruja Region in Albania. *2nd International ISEKI_ FOOD Conference*. 2011a. Milano, Italy.

[19] Topi D, Thomaj F, Halimi E: Virgin Olive Oil production from the major olive varieties in Albania. *Poljoprivreda i Sumarstvo*; 2012b; 58(2):87.

[20] Velo S. and Topi D: Study of *Kalinjot* extra virgin olive oil, fatty acids profiles, and trans-isomers. *Journal of Hygienic and Engineering Design,* 2015;12:129-135.

[21] Velo S. and Topi D: Analytical study of *Kalinjot* olive cultivar in the southern regions of Albania. III International Congress, "Food Technology, Quality, and Safety," 2016. 25-27.

[22] Velo S. and Topi D: Characterization of *Kalinjot* and *Nisiot* Monocultivar Virgin Olive Oils produced in Albania. *Asian Journal of Chemistry;* 2017; 29 (6), 1347-1350.

[23] Angerosa F, Campestre Ch, Giansante L: Analysis and Authentication. In Boskou B. *Olive Oil Chemistry and Technology,* Champaign, USA: AOCS Press; 2006; Vol. 1, pp. 450.

[24] Topi D, Thomaj F, Carvalho AP, Gomes A: Characterization of olive varieties from Albania: Fatty acid profile and total phenolic content. *Food Technologists, Biotechnologists and Nutritionists*; 2011b. 185.

[25] Aparicio R, and Luna. G: Characterization of monovarietal virgin olive oils. *European Journal of Lipid Science and Technology*. 2002; 104: 614–627.

[26] Pinelli P, Galardi C, Mulinacci N, Vincieri FF, Cimato A, Romani A: Minor polar compound and fatty acid analyses in monocultivar virgin olive oils from Tuscany. *Food Chemistry*; 2003; **80**: 331-336.

[27] Haddada M, Krichène Dh, Manai H, Oueslati I, Daoud D, Zarrouk M: Analytical evaluation of six Monovarietal virgin olive oils from Northern Tunisia. *European Journal of Lipid Science and Technology*; 2008; **110** (10):905–913.

[28] Kiritsakis AK, Nauos GG, Polymenoupoulos Z, Thomai T, Sfakiotakis EY: Effect of fruit storage conditions on olive oil quality. *Journal of American Oil Chemists Society*; 1998; **75**: 721-724.

[29] Servili M. Sordini B. Esposto S. Urbani S. Veneziani G. Di Maio I. Selvaggini R. Taticchi A: Biological Activities of Phenolic Compounds of Extra Virgin Olive Oil. *Antioxidants*, 2014;3:1-23. doi:10.3390/ antiox3010001.

[30] De Medina V, Priego-Capote F, de Castro L: Characterization of monovarietal virgin olive oils by phenols profiling. *Talanta*; 2015; 132:424–432.

[31] Del Monaco G, Officioso A, D'Angelo S, La Cara F, Ionata E, Marcolongo L, Squillaci G, Maurelli L, Morana A: Characterization of extra virgin olive oils produced with typical Italian varieties by their phenolic profile. *Food Chemistry*. 2015; 184:220–228.

[32] Seguera-Carretero A, Memendez-Menendez J, Fernandez-Gutierrez A: Polyphenols in Olive Oil: The importance of phenolic compounds in the chemical composition of olive oil. In: In Olives and Olive oil in Health and Disease Prevention. Editors: V. R. Preedy and R. R. Watson. Elsevier. London, UK. 2010; Pp. 167-175.

[33] El Riachy M, Priego-Capote F, Leon L, Rallo L, Luque de Castro MD: Hydrophilic antioxidants of virgin olive oil. Part 1: Hydrophilic phenols: A key factor for virgin olive oil quality. *European Journal of Lipid Science and Technology*. 2011; **113**:678–691

[34] Cicerale S, Conlan X.A, Sinclair AJ. Keast RS: Chemistry and health of olive

oil phenolics. *Critical Reviews in Food Science and Nutrition*. 2009; 49(3): 218-236. DOI: 10.1080/ 10408390701856223.

[35] EFSA Panel on Dietetic Products, Nutrition and Allergies (NDA), *EFSA Journal*, 2011;9(4):2033 (25 pp.). doi: 10.2903/j.efsa.2011.2033. Available online: www.efsa.europa.eu/efsajournal.

[36] Kelebek, H., Kesen, S., & Selli, S. Comparative study of bioactive constituents in Turkish olive oils by LC-ESI/MS/MS. *International Journal of Food Properties*, 2015; **18**(10), 2231-2245.

[37] Montedoro, G, Servilli, M, Miniati, E: Simple and hydrolysable Phenolic Compounds in Virgin Olive oil. 1. Their extraction, separation, and quantitative and semi quantitative evaluation by HPLC. *Journal of Agriculture and Food Chemistry*. 1992;*40*:1571-1576.

[38] Topi D, Guclu G, Kelebek H, Selli S: Comparative elucidation of phenolic compounds in Albanian olive oils using LC-DAD-ESI-MS/MS. *Journal of Liquid Chromatography & Related Technologies;* 2020; 43(5-6):203-212.

[39] Oliveras-Lopez J, Innocenti M, Giaccherini C, Ieri F, Romani, A, Mulinacci N: Study of the phenolic composition of Spanish and Italian monocultivar Extra Virgin Olive Oils: Distribution of lignans, secoiridoidic, simple phenols, and flavonoids. *Talanta*. 2007; **73**:726–732

[40] Bayram B, Esatbeyoglu T, Schulze N, Ozcelik B, Frank J, Rimbach G: Comprehensive Analysis of Polyphenols in 55 Extra Virgin Olive Oils by HPLC-ECD and Their Correlation with Antioxidant Activities. *Plant Food Hum Nutr*. 2012; **67**:326–336

[41] Kulisić-Bilusić T, Melliou E, Giacometti J, Causević A, Corbo S, Landeka M, Magiatis P: Phenolics, fatty acids, and biological potential of selected Croatian EVOOs. *European Journal of Lipid Science and Technology*. 2017; **119**(10):1700108.

[42] Luna, G, Morales, MT, & Aparicio, R: Characterization of 39 varietal virgin olive oils by their volatile compositions. *Food Chemistry*. 2006;*98*:243–252.

[43] Sarolić M, Gugić M, Tuberoso G, Jerković I, Suste M, Marijanović Z, & Kuś M: Volatile Profile, Phytochemicals and Antioxidant Activity of Virgin Olive Oils from Croatian Autochthonous Varieties *Masnjaca* and *Krvavica* in comparison with Italian Variety Leccino. *Molecules*. 2014;*19*:881–895.

[44] Kesen S, Kelebek H, & Selli S: Characterization of the volatile, phenolic and antioxidant properties of monovarietal olive oil obtained from cv. *Halhali*. *Journal of American Oil Chemists' Society*. 2013;*90*:1685–1696.

Chapter 7

Cooking with Extra Virgin Olive Oil

Ana Florencia de Alzaa, Claudia Guillaume and Leandro Ravetti

Abstract

Mediterranean cultures have used Extra Virgin Olive Oil (EVOO) as the only source of cooking oil for centuries, with their diet showing the highest amount of scientifically proven health benefits. However, there is a common misconception that EVOO is not suitable for cooking given its relatively lower smoke point, despite no scientific evidence that support this. This chapter aims to provide an overview of how EVOO is healthier, safer, and more stable to cook with than other common edible oils. Furthermore, this chapter aims to present EVOO's suitability for use on Teflon coated pans, which is another common myth.

Keywords: extra virgin olive oil, cooking oils, smoke point, nutrient content, key minor components, bioactives

1. Introduction

Olive oil occupied a key role in ancient Roman cooking and its culinary use expanded along with the Roman Empire, laying the foundation for what we now refer to as the Mediterranean diet. The culinary use of olive oil takes four basics forms: as a preservative, as a cooking medium, as an ingredient and as a condiment [1]. This chapter has a central focus on EVOO as a cooking medium.

As a cooking medium EVOO has a many functions, such as transferring heat from the heat source to the food, acting as a lubricant to prevent food from sticking to the cooking surface, adding flavour, crust and creating a more visually appealing look to the food [1]. It is an extremely healthy oil to use for all types of cooking and there is a lot of existing and emerging research related to the health benefits due to high levels of antioxidants (some of which are unique to EVOO) and the ability of the oil to enhance the health attributes of some ingredients once cooked [2–4].

However, myths related to cooking with EVOO have been prevalent from time to time, creating a lot of confusion for consumers.

2. Debunking the myths

Some of the prevalent myths related to cooking with EVOO are listed in **Table 1**.

Myth	Assumptions based on the myth	Truth/reality based on scientific evidence
Smoke point is a relevant factor in determining how suitable an oil is to cook with.	EVOO is not suitable for cooking at high temperatures given its lower smoke point.	The utilisation of smoke point as an indicator of the ability of an oil to withstand heat, and to determine suitability for cooking is technically incorrect, and is not supported by scientific evidence. Recent evidence [5] shows that EVOO is the most stable oil when heated when compared to other edible oils with higher smoke points. Mediterranean cultures have used EVOO as their only source of cooking oil for centuries and their diet has the highest amount of scientifically proven health benefits [2, 6, 7].
Cooking with EVOO can ruin cookware, such as non-stick pans (e.g., Teflon coated pans).	EVOO could be damaging to cookware coating.	Although there is no published scientific evidence to support this, these beliefs are specifically supported by some cookware manufacturers' specifications that oils with higher smoke points are more suitable for cooking with Teflon coated cookware [8, 9]. On the contrary, EVOO, like any other oil, acts as a lubricant, preventing the food from sticking to the pan [1]. Cooking with EVOO does not ruin non-stick Teflon coated pans at a different rate than other cooking oils.
Heating olive oil will increase the amount of saturated or trans fats.	You cannot heat olive oil.	All oils will oxidise and hydrogenate to a minor degree when heated several times using high temperatures, such as those used in industrial frying processes [10]. It has been documented that olive oil is less prone to oxidation and hydrogenation when heated than other oils when heated because it is rich in monounsaturated fat [11, 12]. Cooking with EVOO does not produce significant traces of trans fatty acids. In fact, EVOO is less prone to hydrogenation than other vegetable oils.
When you cook vegetables with EVOO, the vegetables lose antioxidants.	EVOO is not suitable to use when cooking vegetables.	This is incorrect. Recent evidence shows that when cooking with EVOO (including deep frying and sautéing), there is a resultant increase in total phenols (antioxidants) in the cooked food (particularly when cooking raw vegetables) [13]. Cooking with EVOO may in fact improve the nutritional properties of the food.

Table 1.
Myths related to cooking with EVOO.

2.1 Performance of edible oils when heated

Chemical reactions such as hydrolysis, oxidation, and polymerisation are prone to occur when edible oils are heated. Heating oils at high temperatures or for long periods of time can generate decomposition products such as free fatty acids (FFAs), alcohols, cyclic compounds, and polymers. Several factors can affect or influence these reactions, such as the type and quality of the oil, the kind of food used in cooking, the time and temperature of cooking and the food/oil ratio. These chemical reactions can affect both the nutritional value and the organoleptic properties of the oil. In addition, some of the products formed through oil decomposition may have adverse effects on human health. Physical changes in oil occur during heating and include increased viscosity, darkening in colour, and increased foaming. At the same time, the smoke point of the oil decreases [14–18].

There are two major properties of cooking oils commonly believed to dictate the behaviour of that oil, and subsequent safety when exposed to high cooking temperatures: smoke point and oxidative stability. While oxidative stability is a reasonable predictor of an oils' ability to withstand heat, initial smoke point has proven to have very little correlation with the oils' stability under heat while cooking, and the formation of polar compounds [5].

2.2 What is smoke point and why is it an unreliable measure of oil performance when heated?

The smoke point is defined as the temperature at which a visible and continuous bluish smoke appears. At this point sufficient volatile compounds, such as FFAs and short chain oxidation products are emerging and evaporating from the oil.

The smoke point of an oil generally increases as the FFA content decreases, and the degree of refinement increases [19, 20].

The smoke point should not be considered a reliable measure of an oil's stability and suitability for cooking for the following reasons:

- The smoke point changes when an oil is heated, therefore it is not the same during the whole cooking process. The smoke point decreases faster when heating oils with a higher polyunsaturated fat content, such as in seed oils, than when heating oils with less poly-unsaturation and greater monounsaturated fat levels such as in EVOO [21].

- The chemical fraction that mostly determines the smoke point of an oil is the FFA fraction which is under 1% of the total oil composition. This means that when the oil reaches the smoke point, only a minor part of it is evaporating and does not indicate the deterioration of the fat itself. In fact, studies have shown that the levels of FFA are not a reliable indication of deterioration of cooking fat [14].

- When determining the smoke point, a small volume of oil is heated using a little brass cup in a confined and dark environment (**Figure 1**). Studies have shown that the smoke point rises when using a bigger container or a larger volume of oil in the presence of air. As a result, when cooking in a kitchen, smoke point temperatures could be greater than the ones that have been reported in the literature [22–24]. Thus, exact smoke point temperatures cannot be given [21, 24].

Figure 1.
Smoke point determination equipment.

- The standard procedure used to establish the smoke point relies heavily on the ability of the worker to determine visually the point at which the oil begins to smoke. This means that there can be analyst subjectivity when using this test procedure [19, 24].

- The refining process used to produce other vegetable oils such as canola, peanut and rice bran oils involves high temperatures to neutralise FFAs. This gives the oils a higher smoke point but also produces secondary oxidation products which have been shown to have a detrimental effect on human health.

2.2.1 What measurements we should use instead to determine oil suitability for cooking

When heating oils, the process of fat oxidation is accelerated. Fat oxidation is where fat molecules interact with oxygen, leading to the potential formation of harmful compounds. Many authors agree that oxidative stability is the best predictor of the behaviour of oil during cooking [5, 16, 25].

Industrially and technically, the ability of an oil to withstand heat is measured by its resistance to the formation of polar compounds.

Non-volatile polar compounds, triacylglycerol (TAG) dimers and polymers are the main deterioration products of cooking oils. Several studies have associated these substances with certain types of cancer and neurodegenerative diseases including Alzheimer's and Parkinson's disease [26, 27]. However, this negative effect on health is related to the dose of these components. National and International legislation has identified that no more than 24–27% of polar materials in the final oil is a safe limit for human consumption [27]. These limits are made to ensure the oils used in cooking operations are safe for human consumption. When storing food after being cooked is required, then the recommended end point of polar compounds in the oil is <10% [14, 27].

Laboratory research is extremely important to understand the frying process and what the toxicological limits are to establish guidelines for consumers. However, it is important to highlight that sometimes the research completed with cooking oils has been pushed to the point of abusing the oils. These are extreme heating conditions and the results obtained do not always represent real-life situations, especially during home cooking. Therefore, proper judgement when drawing conclusions needs to be made. When cooking at home, normally one does not reach excessive temperatures for prolonged periods of time that deteriorate either the food or the oil in a way that makes them inedible or unsafe [27].

2.3 Why should we cook with EVOO?

2.3.1 EVOO composition

Edible oils are composed of triacylglycerols (> 96%) and endogenous minor components. It is generally agreed that the inherent composition of edible oils exerts considerable influence on their heating stability [15, 18]. In the interest of understanding better why we can, and we should cook with EVOO, it is important to review first its chemical composition. EVOO has a high level of oxidative stability when compared to other types of cooking oil and is less likely to undergo oxidation. This is primarily attributed to the following factors:

1. EVOO contains high levels of oleic acid, a monounsaturated fatty acid (MUFA) with just one double bond and low levels of linoleic and linolenic acids, which

are polyunsaturated fatty acids (PUFAs) with multiple double bonds. This high MUFA/PUFA ratio confers good oxidative stability making it highly resistant to the production of harmful substances (such as polar compounds). The high levels of double bonds in PUFAs in seed oils make them sensitive to damage by heat [28–30].

2. The presence of natural antioxidants in EVOO, such as biophenols and Vitamin E, enhance the oil's stability and resistance to oxidative degradation. Besides adding health benefits to the oil, these minor constituents boost EVOO's stability reducing oxidative processes when the oil is heated [31, 32].

3. The phytosterols are a significant fraction of the unsaponifiable matter in EVOO. These substances add to the oil's health profile. Some sterols have been shown to provide higher protection against lipid thermal deterioration by decreasing the production of TAG polymers [33]. These sterols are sometimes lost during oil refining and because EVOO does not require this manufacturing step, it retains high concentrations of phytosterols. It has been documented that phytosterols can be transferred to food while cooking, which could have a nutritionally positive impact on consumers [34].

2.3.2 EVOO performance when heated in comparison with other edible oils

In a comprehensive trial in Australia [5] conducted by an ISO 17025 accredited laboratory in 2018, ten of the most used cooking oils were selected from the supermarket to test their performance when heated. The oils tested were EVOO, virgin olive oil (VOO), olive, canola, rice bran, grapeseed, coconut, peanut, sunflower, and avocado oils.

Two different volumes of oils were heated in open pans (250 mL) for 20 minutes from 25–240°C and in deep fryers (3000 mL) at 180°C for 6 hours. Samples were collected at different intervals and then tested.

Authors specifically assessed the correlation between smoke point and other key chemical parameters related to an oils stability and likelihood to break down and form harmful compounds.

From this study, it was concluded that under different heating conditions, the generation of polar compounds with temperature and time was more pronounced for refined seed oils with higher initial values of smoke point, PUFAs, K232 and K270 (oxidative by-products). Reasonable predictors of how an oil will perform when heated have been oxidative stability, secondary products of oxidation, total level of PUFAs. EVOO was the most stable oil of those tested when heated, followed closely by coconut oil and other virgin oils such as avocado and high oleic acid seed oils. EVOO yielded lower levels of polar compounds and TFAs when compared with other oils.

This research also showed that an oil's smoke point is not a relevant parameter to explain the oil's behaviour when heated. Smoke point does not correlate with the stability of the oil during heating, as it showed a positive correlation with the increase in polar compounds (**Table 2**). That is to say that the higher the smoke point, the more polar compounds that are produced. PUFAs, K232 and K270 showed a positive correlation with polar compounds. Oxidative stability was negatively correlated with final content of polar compounds, demonstrating that a non-stable oil in terms of thermal degradation, will produce more polar compounds when heated (**Table 2**).

These results are also supported by recent research carried out in New Zealand in 2019 [25]. The authors concluded that quality EVOO, in accordance with relevant

Initial Parameter	Correlation with final polar compounds levels (%)
Smoke Point	83
Oxidative Stability	−65
FFA	−34
PUFAs	74
UV Coefficient K232	80
UV Coefficient K270	54

Note: A negative, or inverse correlation, between two variables, indicates that one variable increases while the other decreases, and vice-versa. i.e. the less oxidative stability, the more polar compounds produced.
Table Reference [5].

Table 2.
Correlation between final polar compounds and initial oil's chemical parameters.

olive oil standards, is the best cooking oil for use in the home from a stability and health viewpoint. These authors also recommended criteria to indicate an EVOO is stable for cooking:

- Initial % free fatty acids (FFA) <0.2% (w/w as oleic acid)
- Peroxide Value (PV) < 5.0 mEq/kg
- Induction time in Rancimat >15 hours.
- Total polar compounds after 8 hours heating at 180°C < 25%
- p-Anisidine value after 8 hours at 180°C < 70

2.3.3 What are trans fat and why does cooking with EVOO not produce significant TFAs?

TFAs are formed during partial hydrogenation of oils. The interconversion from cis to trans requires a lot of energy (~65 kcal/mole), however the use of a high temperature or a catalyst can enhance the reaction [35]. Consumption of diets high in hydrogenated fat and/or TFAs has been shown to have an adverse effect on lipoprotein profiles with respect to cardiovascular disease risk [36, 37].

The formation of TFAs while cooking food using oil is closely related to the temperature and how many times oil is reused [38, 39]. Several European countries have determined that the frying oil temperature must not exceed 180°C. These measures not only contribute to decreased degradation of unsaturated fatty acids but also result in a lower formation of monounsaturated trans fatty acid (MTFAs) and polyunsaturated trans fatty acids (PTFAs) during frying.

Much research has been done to determine how typical cooking procedures used in food preparation affect TFAs formation in edible oils. Research suggests even applying normal and/or extreme temperatures when cooking does not significantly affect the amounts of TFAs in edible oils [40, 41]. Formation of minor amounts of trans-oleic acid, inferior to 0.2 g/100 g fatty acids was observed by [11, 12] for all the olive oil grades, which is lower than the trans amounts in other refined vegetable oils.

Recent research presented at the World Congress of Oils and Fats in 2020 [42] demonstrated that initially EVOO does not contain TFA and that the food TFA content decreased by approx. 70% or remained stable when using EVOO. The same behaviour was observed with oils: the lowest TFAs production was in EVOO in comparison with other vegetable oils.

2.3.4 Cooking with EVOO

Cooking with edible oils, such as deep frying, usually involves two phenomena. Firstly, when the oil, that acts as a heating medium to the food, reaches 100°C water starts to evaporate from the food. This in turn gives way to the oil being absorbed into the food which modifies the fatty acid composition of the food as it cooks. It has been proven that the fat content of the food after deep frying is more like the fat profile of the oil used to cook than the raw food itself [14]. In addition, although the antioxidant content is reduced somewhat during cooking many healthy substances still remain in EVOO and are absorbed by the food. The absorption of these antioxidants into the food gives the food a better nutritional profile. For this reason, the use of EVOO is a healthier option than using other oils with less bioactive components [13, 42, 43].

2.3.4.1 Frying

Frying is one of the oldest methods of food preparation. It improves the sensory quality of food by formation of aromatic compounds, attractive colour, crust and texture, which are all highly appreciated by consumers [44, 45]. The most common frying methods are deep-frying, being the food totally immersed in hot oil, and pan-frying, when the food is cooked in a pan with a little amount of oil [46, 47].

There is a higher degradation under pan-frying conditions for olive oil and other vegetable oils, that can be explained by the higher contact surface between the food and the oil, higher exposure to atmospheric oxygen, and lower temperature control under processing [46].

Frying with EVOO using a lower food:oil ratio presents lower total polar compound amounts than more unsaturated vegetable oils, and with apparently no interference by the presence of food [32, 48, 49]. Within olive oils, the higher the degree of polyunsaturation the higher the tendency for the formation of total polar compounds [50].

The volatile fraction formed during the heating process, apart from being important from the sensorial point of view, is rich in degradation compounds. The formation of low molecular weight volatile aldehydes has a clear dependence on the temperatures used, rather than frying time [51]. The high oleic acid content in olive oil, together with the presence of chlorophylls, pheophytins and carotenoids, seems to contribute to a reduced acrolein formation and lower amounts of toxic monoaromatic hydrocarbons, alkylbenzenes and alkenylbenzenes, in comparison with other vegetable oils with higher polyunsaturated acyl groups [51–53].

Furthermore, as mentioned previously, in comparison with other vegetable oils, the fried food is enriched with olive oil antioxidants, which improves the nutritional profile of the food [54].

2.3.4.2 Roasting

Roasting with olive oil is common in both domestic and industrial food preparation in Mediterranean countries [55]. This procedure is highly prone to oxidation due to the higher surface area exposed to convention hot air and processing times. When comparing with other vegetable oils with a higher degree of unsaturation, olive oil is also more resistant to oxidation under these heating conditions [11, 56].

In opposition, the total polar compounds clearly increase with vegetable oils with higher unsaturation degrees such as sunflower and corn oil [57].

2.3.4.3 Microwave

In general, heating olive oil using a microwave demonstrates an apparent higher oxidation when compared with conventional heating, despite being probably lower than those achieved with other vegetable oils [58]. Researchers have compared microwave and conventional heating (in an electric oven) in several vegetable oils including sunflower, high oleic sunflower and olive oil. Among the studied oils, the EVOO exhibited better performance against oxidation with both heating methods. This is mainly due to its composition, including minor compounds with antioxidant properties (phenolic compounds and tocopherols) and a lower percentage of linoleic acid [59]. Still, all studies were performed without the presence of food, meaning further studies using real processing conditions are required for correct inferences [15].

2.3.5 EVOO and cookware interaction

In 2019, Modern Olives Laboratory, an Australian oil specialist laboratory, conducted research to assess the suitability of various cooking oils, including EVOO, for use on Teflon coated (TC) pans.

To investigate the hypothesis of whether cooking with EVOO ruins pans, the researchers measured the release of elements and metals from the pans when separately heated with different oils. They used three different brands of TC pans. These pans were heated with an acidic solution of water vinegar (WV) both prior to and after 6 cycles of heating with different oils (EVOO, olive oil, canola oil, rice bran oil and grapeseed oil). The WV solutions were tested to study the release of various metals.

Combining all TC pans, the authors found no significant differences in the chemical elements content between the final WV solutions from TC pans treated with the different oils. This indicates there is no significant difference between the volume of metals released from the cookware when various cooking oils were used. Hence, the various cooking oils had no effect on the pans' integrity and quality when cooking. However, differences of statistical significance for Ca, Cu, Fe, P, Zn and SiO2 were observed between the different TC pan types. Higher values of these metals were detected in the most expensive pan compared with the cheapest TC pan. For example, Ca average values (including initial and final treatment) in the most expensive pan were ~ 2.92 mg/L vs. ~1.75 mg/L in the cheapest pan and ~ 2.42 mg/L in the average price pan. When considering each brand of TC pan, phosphorus levels were significantly higher between treatments when using rice bran oil in the average priced TC pan (4.7 mg/L vs. 2.5 mg/L) versus a low- or high-priced TC pan. Silicon dioxide was not detected before treatment and significantly increased using olive (1.1 mg/L) and grapeseed (1.03 mg/L) oils only in the lowest priced TC pan.

After all treatments, no visual deterioration of any of the TC pans was observed. This investigation indicates that higher differences in metal leaching were between pans quality, rather than between the treatments with the different oils. In no case the use of EVOO lead to the release of significantly higher levels of metallic substances from the pan than when using any other oil.

Even though these results are limited considering the lifetime of the TC pan, they indicate no initial impact of the oils' smoke point on the performance of the TC pan and that EVOO performs similarly to other oils under normal cooking conditions when it comes to TC pan degradation.

3. Conclusion

Sufficient research has been done to demonstrate that an oil's smoke point is not a reliable measurement as an indicator of the ability of an oil to withstand heat, and to determine suitability for cooking. Reasonable predictors of how an oil will perform when heated are oxidative stability, secondary products of oxidation, and total level of PUFAs. EVOO has been demonstrated to be the most stable oil when heated given its unique chemical composition, which is rich in monounsaturated fatty acid and antioxidant content.

Experts have agreed that one of the most versatile and healthy oils to cook with is EVOO and many studies have linked it to better heart and overall health.

Food cooked with EVOO also had lower levels of undesirable products of degradation such as TFAs and polar compounds when compared with other vegetable oils such as canola, grapeseed, peanut, sunflower and rice bran oils, while deep-frying under normal cooking conditions.

Furthermore, based on scientific evidence EVOO does not deteriorate the coating when using Teflon cookware. On the contrary it acts as a lubricant to prevent food from sticking to the pans.

Selecting a true high quality EVOO, that is certified to meet EVOO grade and quality requirements is important to ensure high oxidative stability and safety while cooking.

Acknowledgements

We thank the Olive Wellness Institute (OWI) and Modern Olives Laboratory for their help during this work. To know more about OWI please visit https://olivewellnessinstitute.org/.

Conflict of interest

The authors declare no conflict of interest.

Author details

Ana Florencia de Alzaa*, Claudia Guillaume and Leandro Ravetti
Modern Olives, Lara, Australia

*Address all correspondence to: f.dealzaa@modernolives.com.au

References

[1] Tardi A. The Extra-Virgin Olive Oil Handbook2014. 321-37 p.

[2] Ramirez-Anaya Jdel P, Samaniego-Sanchez C, Castaneda-Saucedo MC, Villalon-Mir M, de la Serrana HL. Phenols and the antioxidant capacity of Mediterranean vegetables prepared with extra virgin olive oil using different domestic cooking techniques. Food Chem. 2015;188:430-8.

[3] Perez-Herrera A, Rangel-Zuñiga OA, Delgado-Lista J, Marin C, Perez-Martinez P, Tasset I, et al. The antioxidants in oils heated at frying temperature, whether natural or added, could protect against postprandial oxidative stress in obese people. Food chemistry. 2013;138(4):2250-9.

[4] Gray S. Cooking with extra virgin olive oil. Journal of the Australasian College of Nutritional and Environmental Medicine. 2015; 34(2):8.

[5] De Alzaa F, Guillaume C, Ravetti L. Evaluation of Chemical and Physical Changes in Different Commercial Oils during Heating. ACTA Scientific Nutritional Health. 2018;2(6):2-11.

[6] Assaf-Balut C, Garcia de la Torre N, Duran A, Fuentes M, Bordiu E, Del Valle L, et al. A Mediterranean diet with additional extra virgin olive oil and pistachios reduces the incidence of gestational diabetes mellitus (GDM): A randomized controlled trial: The St. Carlos GDM prevention study. PLoS One. 2017;12(10):e0185873.

[7] Yaqoobb P. Nutritional and health aspects of olive oil. Eur J Lipid Sci Technol. 2002;104:685-97.

[8] Woll. Manufacturer booklet [Available from: https://www.woll-cookware.com/en/use-care/

[9] Diamond® S. Manufacturer Booklet [Available from: https://swissdiamond.wordpress.com/2012/09/26/choosing-a-cooking-oil/ https://swissdiamond.wordpress.com/tag/cleaning/

[10] Kiritsakis AK. Olive oil: from the tree to the table: Food & Nutrition Press; 1998.

[11] Albi T, Lanzón A, Guinda A, León M, Pérez-Camino MC. Microwave and Conventional Heating Effects on Thermoxidative Degradation of Edible Fats. Journal of Agricultural and Food Chemistry. 1997;45(10):3795-8.

[12] Caponio F, Pasqualone A, Gomes T. Changes in the fatty acid composition of vegetable oils in model doughs submitted to conventional or microwave heating. International Journal of Food Science & Technology. 2003;38(4):481-6.

[13] Ramírez-Anaya JP, Samaniego-Sánchez C, Castañeda-Saucedo M, Villalón-Mir M, de la Serrana H. Phenols and the antioxidant capacity of Mediterranean vegetables prepared with extra virgin olive oil using different domestic cooking techniques. Food chemistry. 2015;188:430-8.

[14] WARNER K. Chemical and physical reactions in oil during frying. In: GUPTA MKW, K.; WHITE, P. J. , editor. Frying Technology and Practices USA: AOCS Press; 2004. p. 16-28.

[15] Santos CS, Cruz R, Cunha SC, Casal S. Effect of cooking on olive oil quality attributes. Food research international. 2013;54(2):2016-24.

[16] Aladedunye FA. Curbing thermo-oxidative degradation of frying oils: current knowledge and challenges. European Journal of Lipid Science and Technology 2015;117(11):1867-81.

[17] Zribi A, Jabeur H, Aladedunye F, Rebai A, Matthäus B, Bouaziz M. Monitoring of Quality and Stability Characteristics and Fatty Acid Compositions of Refined Olive and Seed Oils during Repeated Pan- and Deep-Frying Using GC, FT-NIRS, and Chemometrics. Journal of Agricultural and Food Chemistry. 2014;62(42):10357-67.

[18] Shahidi F, Zhong Y. Lipid oxidation and improving the oxidative stability. Chemical Society Reviews. 2010;39(11):4067-79.

[19] AOCS. Official Method Cc 9a-48. Smoke, Flash, and Fire Points, Cleveland Open Cup Method: the American Oil Chemists' Society (AOCS).

[20] Nielsen S. Food Analysis. 4 ed: Springer US; 2010. 602 p.

[21] Wolke RL. What Einstein Told His Cook: Kitchen Science Explained2002. 79-81 p.

[22] Servet Gulum; Sumnu SS. Advances in Deep-Fat Frying Foods. 1st ed: CRC Press; 2009. 321 p.

[23] Blumenthal M. A new look at the chemistry and physics of deep-fat frying. Food technology (Chicago). 1991;45(2):68-71.

[24] NAOOA. Understanding cooking oil smoke points 2016 [Available from: https://www.aboutoliveoil.org/understanding-cooking-oil-smoke-points#:~:text=The%20surface%20area%20also%20has,smoke%20at%20home%20while%20cooking.

[25] Eyres L, Wong M. Food New Zealand 12. 2019.

[26] O'Brien PJ, Siraki AG, Shangari N. Aldehyde sources, metabolism, molecular toxicity mechanisms, and possible effects on human health. Critical reviews in toxicology. 2005;35(7):609-62.

[27] Stier RF. Toxicology of frying fats and oils. Frying Technology and Practices2004. p. 178-94.

[28] Warner K, Frankel E, Mounts T. Flavor and oxidative stability of soybean, sunflower and low erucic acid rapeseed oils. Journal of the American Oil Chemist" Society. 1989;66:558-64.

[29] Parker TD, Adams DA, Zhou K, Harris M, Yu L. Fatty Acid Composition and Oxidative Stability of Cold-pressed Edible Seed Oils. Journal of Food Science. 2003;68(4):1240-3.

[30] Shahidi F. Oxidative Stability of Edible Oils as Affected by Their Fatty Acid Composition and Minor Constituents. ACS Symposium Series. 2002;836:201-11.

[31] Akil E, Castelo-Branco VN, Costa AMM, do Amaral Vendramini AL, Calado V, Torres AG. Oxidative stability and changes in chemical composition of extra virgin olive oils after short-term deep-frying of french fries. Journal of the American Oil Chemists' Society. 2015;92(3):409-21.

[32] Casal S, Malheiro R, Sendas A, Oliveira BP, Pereira JA. Olive oil stability under deep-frying conditions. Food and Chemical Toxicology. 2010;48(10):2972-9.

[33] Singh A. Sitosterol as an antioxidant in frying oils. Food Chemistry. 2013;137(1):62-7.

[34] Salta F, Kalogeropoulos N, Karavanou N, Andrikopoulos N. Distribution and retention of phytosterols in frying oils and fried potatoes during repeated deep and pan frying. European Food Research and Technology. 2008;227:391-400.

[35] Kodali DR. Trans Fats - Chemistry, occurence, functional need in foods and potential solutions. Trans fats alternatives. USA: AOCS; 2006. p. 4.

[36] Ganguly R, Pierce GN. The toxicity of dietary trans fat",. Food and Chemical Toxicology. 2015:78. 170-6.

[37] Lopez-Garcia E, Schulze MB, Meigs JB, Manson JE, Rifai N, Stampfer MJ, et al. Consumption of trans fatty acids is related to plasma biomarkers of inflammation and endothelial dysfunction. Journal of Nutrition. 2005:135, 562-6.

[38] Moreno M, Olivares D, Lopez F. Determination of unsaturation grade and trans isomers generated during thermal oxidation of edible oils and fats by FTIR. Molec Struct. 1999:482: 551-6.

[39] Sanibal E, Mancini-Filho J. Perfil de ácidos graxos trans de óleo e gordura hidrogenada de soja no processo de fritura. Ci Tecnol Alim. 2004:24: 7-31.

[40] Song J, Park J, Jung J, Lee C, Gim SY, Ka H, et al. Analysis of Trans Fat in Edible Oils with Cooking Process. Toxicol Res. 2015;31(3):307-12.

[41] Przybylski O, Aladedunye FA. Formation of trans fats during food preparation. Canadian journal of dietetic practice and research : a publication of Dietitians of Canada = Revue canadienne de la pratique et de la recherche en dietetique : une publication des Dietetistes du Canada. 2012;73(2):98-101.

[42] de Alzaa F, Guillaume C, Ravetti L. Preliminary Results for the Evaluation of Chemical and Nutritional Changes in Chips, Chicken Nuggets and Broccoli after Deep Frying with Extra Virgin Olive Oil, Canola and Grapeseed Oils. Australia: Modern Olives Laboratory Services,; 2020.

[43] Moreno DA, López-Berenguer C, García-Viguera C. Effects of Stir-Fry Cooking with Different Edible Oils on the Phytochemical Composition of Broccoli. Journal of food science. 2007;72(1).

[44] Echarte M, Ansorena D, Astiasarán I. Consequences of microwave heating and frying on the lipid fraction of chicken and beef patties. J Agric Food Chem. 2003;51(20):5941-5.

[45] Pedreschi F. Frying of Potatoes: Physical, Chemical, and Microstructural Changes. Drying Technology. 2012;30(7):707-25.

[46] Nikolaos A, Nick K, Angeliki F, Maria B. Performance of virgin olive oil and vegetable shortening during domestic deep-frying and pan-frying of potatoes. International Journal of Food Science and Technology 2002;37:177-90.

[47] Sioen I, Haak L, Raes K, Hermans C, De Henauw S, De Smet S, et al. Effects of pan-frying in margarine and olive oil on the fatty acid composition of cod and salmon. Food Chemistry. 2006;98(4):609-17.

[48] Bastida S, Sánchez-Muniz FJ. Polar content vs. TAG oligomer content in the frying-life assessment of monounsaturated and polyunsaturated oils used in deep-frying. Journal of the American Oil Chemists' Society. 2002;79(5):447-51.

[49] Kalogianni EP, Karastogiannidou C, Karapantsios TD. Effect of potato presence on the degradation of extra virgin olive oil during frying. International Journal of Food Science & Technology. 2010;45(4):765-75.

[50] Brenes M, García A, Dobarganes MC, Velasco J, Romero C. Influence of Thermal Treatments Simulating Cooking Processes on the Polyphenol Content in Virgin Olive Oil.

Journal of Agricultural and Food Chemistry. 2002;50(21):5962-7.

[51] Fullana A, Carbonell-Barrachina AA, Sidhu S. Comparison of volatile aldehydes present in the cooking fumes of extra virgin olive, olive, and canola oils. J Agric Food Chem. 2004;52(16):5207-14.

[52] Procida G, Cichelli A, Compagnone D, Maggio RM, Cerretani L, Del Carlo M. Influence of chemical composition of olive oil on the development of volatile compounds during frying. European Food Research and Technology. 2009;230(2):217-29.

[53] Uriarte PS, Guillén MD. Formation of toxic alkylbenzenes in edible oils submitted to frying temperature: Influence of oil composition in main components and heating time. Food Research International. 2010;43(8):2161-70.

[54] Kalogeropoulos N, Mylona A, Chiou A, Ioannou M, Andrikopoulos N. Retention and distribution of natural antioxidants (α-tocopherol, polyphenols and terpenic acids) after shallow frying of vegetables in virgin olive oil. LWT - Food Science and Technology. 2007;40: 1008-17.

[55] Silva LM, Pinto J, Carrola J, Paiva-Martins F. Oxidative stability of olive oil after food processing and comparison with other vegetable oils. Food Chemistry. 2010;121:1177-87.

[56] Caponio F, Pasqualone A, Gomes T. Effects of conventional and microwave heating on the degradation of olive oil. European Food Research and Technology. 2002;215(2):114-7.

[57] Silva L, Garcia B, Paiva-Martins F. Oxidative stability of olive oil and its polyphenolic compounds after boiling vegetable process. LWT - Food Science and Technology. 2010;43(9):1336-44.

[58] Malheiro R, Casal S, Ramalhosa E, Pereira JA. Microwave Heating: A Time Saving Technology or a Way to Induce Vegetable Oils Oxidation? Advances in Induction and Microwave Heating of Mineral and Organic Materials2011.

[59] Albi T, Lanzón A, Guinda A, Pérez-Camino MC, León M. Microwave and Conventional Heating Effects on Some Physical and Chemical Parameters of Edible Fats. Journal of Agricultural and Food Chemistry. 1997;45(8):3000-3.

Chapter 8

Virgin Olive Oil Phenolic Compounds: Insights on Their Occurrence, Health-Promoting Properties and Bioavailability

El Amine Ajal, Salah Chaji, Sanae Moussafir, Rachid Nejjari, Abdelmajid Soulaymani and Aadil Bajoub

Abstract

Virgin olive oil is a highly appreciated edible oil, considered as a relevant component of the Mediterranean diet. The spread of this foodstuff all over the world is making, to a certain extent, that new markets and consumers are getting used to this "Mediterranean's golden treasure". Currently, there is great momentum in research relating virgin olive oil intake to healthiness, which has been mainly associated with its phenolics content. Phenolics are considered health-promoting compounds due to their multifaceted biochemical actions that can potentially reduce the risk of various health problems. Yet, since the health-promoting effects of various phenolic compounds have been widely attributed to their metabolic products rather than the naturally occurring forms, the assessment of virgin olive oil phenolics bioavailability is still gaining immense attention and considered a great hot topic among researchers. In the first section of this contribution, the main groups of phenolic compounds identified in virgin olive oil are described, their qualitative and quantitative variability is discussed while analytical approaches applied for their determination are highlighted. The second section reports the beneficial health properties of virgin olive oil consumption related to its phenolics content paying special attention to their bioavailability.

Keywords: Virgin olive oil, phenolic compounds, bioavailability, health-promoting proprieties, Mediterranean diet

1. Introduction

For centuries, virgin olive oil represented the foremost source of lipid in the daily cuisine of most populations around the Mediterranean Basin owing to its unique sensory characteristics that naturally flavour dishes. Nowadays, even this vegetable oil is still being extensively appreciated for its organoleptic attributes, being consumed either fresh or as flavouring ingredient in prepared foods; nevertheless, its consumption has gained enormous significance worldwide due to its well-established health benefits. Indeed, over the last few decades, the phytochemical profile and health-promoting properties of virgin olive oil have been extensively explored. There are several studies in which scientists suggest that virgin olive oil intake contributes to improve human's health and well-being by providing protective effects against a plethora of chronic and cardiovascular diseases, neurodegenerative and ageing-related degenerative disorders. Within the frame of virgin olive oil's healthy properties investigation, over the past decades, two main research lines have been proposed. In one of them, the stress has been placed on evaluating and proving biological features and bioavailability of several of its bioactive compounds, particularly, phenolics [1]. In the second one, several epidemiological studies and clinical trials (such as EUROLIVE, Predimed and European Prospective Investigation into Cancer and Nutrition (EPIC)) have brought to the fore the effects of regular intake of virgin olive oil on health [2–4]. Further detailed information about this matter can be found in remarkably interesting manuscripts, which give a deep insight into recent clinical studies, showing the effects of dietary virgin olive oil intake on the human health [5, 6]. Likewise, very stimulating information about the main bioactive compounds which naturally occur in virgin olive oil and their health effects are well detailed in various relevant publications [7, 8].

Even though the nutritional value of virgin olive oil has been attributed to its overall composition involving various chemical substances such as fatty acids, aliphatic and triterpenic alcohols, phytosterols, tocopherols and phenolic compounds, the latter are widely accepted to be the main contributors. Particularly, the bioactivity exerted by these compounds has been associated with their antioxidant activities as indicated by several studies that identified, through *in vitro* and *in vivo* experiments, a strong relationship between the level of virgin olive oil phenolics content and oxygen radical absorbance capacity, free radical scavenging and ferric reducing ability [9]. This propriety is mainly due to the presence in their chemical structure of one or multiple hydroxyl groups able to donate electrons or hydrogen atoms neutralising, in this manner, free radicals and other reactive oxygen species which allows them to act as reducing agents and singlet oxygen quenchers [10].

Moreover, whereas the phenolic profile of virgin olive oil depends widely on the cultivated variety, orchard geographical location (edaphoclimatic conditions), olive growing system and cultivation practices, harvesting time, processing technologies and conservation conditions; however, data accrued from both clinical and experimental studies have conclusively shown that the health advantages of these bioactive compounds are based on their bioavailability considered, thereby, as the main precondition for their efficient biological effect [11, 12]. Additionally, since the most abundant phenolic compounds in this foodstuff are not necessarily those who exert the highest biological effects, understanding the bioavailability of each one of these metabolites is thereby of utmost importance to establish convincing evidence for their efficiency in health improvement and disease prevention. Having that in mind, it is clear that collecting as much as possible data about the bioavailability of the ingested virgin olive oil phenolic compounds is essential to understand their real biological effects.

This chapter brings new insights on the phenolic compounds as one of the main virgin olive oil bioactive components. Major attention has been paid on phenolic fraction composition, the genotypic and agro-environmental and technological factors implicated, analytical approaches for its determination, and their potential health benefits and bioavailability.

2. Phenolics: virgin olive oil bioactive fraction's key compounds

The positively and highly correlated relationship between virgin olive oil intake and health has increased consumer's demand for more information related particularly to its content on bioactive metabolites such as phenolic compounds. Virgin olive oil is characterised by a wide diversity of phenolics and high total phenolic content ranging from 110 to 900 mg caffeic acid equivalents per kg depending on varietal and geographical origins among other factors [13]. Likewise, it is widely recognised that the phenolics content of virgin olive oil strongly depends on the initial concentration of these compounds in olive fruit. It was established that these metabolites are present in concentration ranges between 1% and 3% of the weight of the fresh pulp [10]. However, given their hydrophilic nature and their partition coefficient between aqueous and oil phases during oil processing, a large amount of these antioxidants is lost with the olive by-products (olive mill wastewater and pomace). Consequently, only 1–2% of the phenolic amount naturally present in the olive fruit could be found in virgin olive oil [14]. Another key characteristic of these metabolites lies in their chemical structures largely different from those identified in corresponding olive fruit. Indeed, the latter shows a complex and highly diversified phenolic composition where glycosylated forms of oleuropein and ligstroside are the most abundant [15]. They are believed to be the precursors of the main secoiridoids compounds found in virgin olive oil [16]. In this regard, during the mechanical extraction of virgin olive oil (particularly the crushing and malaxation steps), various enzymatic and nonenzymatic hydrolysis and oxidation reactions take place inducing several transformative changes in the chemical structure of olive fruit native phenols resulting on the generation of new phenol derivatives [17, 18]. For instance, during the crushing and malaxation steps some enzymatic reactions occur when enzymes (such as polyphenol oxidase, peroxidase, and lipoxygenase) and glycosylated forms of oleuropein and ligstroside meet. Among these enzymes, β-glucosidase seems to play a significant role in the transformation of these compounds into various secoiridoids derivatives such as dialdehydic forms of decarboxymethyl elenolic acid esterified with tyrosol and hydroxytyrosol (oleocanthal (*p*-hydroxyphenyl-ethanol linked to dialdehydic form of elenolic acid (*p*-HPEA-EDA)), and oleacein (3,4-dihydroxyphenyl-ethanol linked to dialdehydic form of elenolic acid (3,4-DHPEA-EDA)) [19]. This fact may explains the rapid decrease of oleuropein and ligstroside concentrations when passing from olive fruit to the corresponding oil [20]. Likewise, hydrolytic mechanisms are known to be involved in the release of hydroxytyrosol and tyrosol in virgin olive oil during storage from complex secoiridoids [21]. Totally, we conclude that phenolic compounds in olive fruits are predominantly found in their glycosylated form whereas conjugate-free compounds, known as aglycones, are the most abundant ones in the corresponding virgin olive oil.

Keeping in mind the above-mentioned, we can claim that virgin olive oil phenolic fraction encompasses those metabolites originally present in olive fruit and those generated during processing.

2.1 Classification of virgin olive oil phenolic compounds

Being abundant in the *Oleaceae* family, which includes *Olea europaea* L., phenolic compounds are a large group of secondary metabolites composed of an aromatic heterocyclic ring bearing one or more hydroxyl groups. The presence of aromatic ring and hydrogen atom of phenolic hydroxyl group makes them as weak acids. To date, more than 32 phenolic compounds have been isolated and identified from virgin olive oil samples using various analytical approaches [22]. For instance, the key phenolic compounds in this vegetable oil can be classified into different chemical classes according to their chemical structures considering mainly the number of aromatic rings, the elements that bind the rings with each other, and the substituents linked to the rings. Accordingly, six chemical families have been reported present in virgin olive oil; namely: simple phenols, secoiridoids, phenolic acids, lignans, flavonoids and hydroxy-isocromans. A representative structure of the main phenolic compounds identified in this product are illustrated in **Figure 1**.

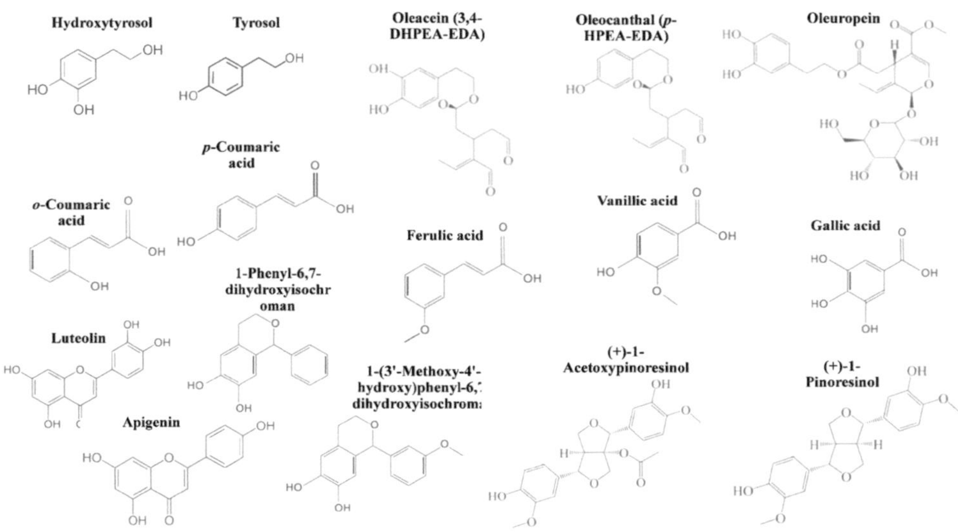

Figure 1.
Chemical structure of the main phenolic compounds found in virgin olive oil.

In a previous metabolomic study of phenolic fraction of virgin olive oil samples obtained from various Mediterranean varieties cultivated in Morocco, a comprehensive analysis was performed by liquid chromatography coupled to mass spectrometry (LC–MS) suggesting that even quantitative differences were observed, the phenolic profiles did not show a significant qualitative differences between studied cultivars [22]. Furthermore, results revealed that secoiridoids prevailed in all analysed samples. A similar result about the phenolic composition of the main cultivated varieties worldwide has also been reported by various authors [23].

2.1.1 Simple phenols

The main phenolic alcohols identified in virgin olive oil are hydroxytyrosol (3,4-dihydroxyphenyl-ethanol (3,4-DHPEA)) and tyrosol (*p*-hydroxyphenyl-ethanol (*p*-HPEA)). Their concentrations are usually low in fresh samples but increases with advanced storage time in a proportional manner due to the hydrolysis of secoiridoids (oleuropein and ligstroside aglycones) [24]. Whilst tyrosol, which has one single hydroxyl group substitution, has been related to a weak antioxidant ability, hydroxytyrosol displays a great radical-scavenging power leading to a better prevention from cardiovascular diseases and to plethora of health effects including anti-inflammatory, anti-cancer and anti-age activities [25]. Thus, in view of the health-promoting effects of this compound and its derivatives, recently, European Food Safety Authority (EFSA) approved a health claim stating that the dietary intake of olive oil phenolic compounds could be able to prevent low density lipoprotein (LDL) oxidation. The exact wording of the claim is olive oil polyphenols contribute to the protection of blood lipids from oxidative stress. The EU restricts the use of this claim to olive oils which contains at least 5 mg of hydroxytyrosol and its derivatives per 20 g of olive oil [26].

2.1.2 Secoiridoids

In virgin olive oil, secoiridoids derivatives are the major group of phenolic compounds. Oleocanthal, and oleacein are among the most abundant ones [16]. These secoiridoids together with the aglycon forms of oleuropein

(3,4-dihydroxyphenyl-ethanol linked to elenolic acid (3,4-DHPEA-EA)) and ligstroside (*p*-hydroxyphenyl-ethanol linked to elenolic acid (*p*-HPEA-EA)) have been associated with some remarkable health effects of virgin olive oil intake [27]. Other derivatives can also be found in lower amount, mainly the aldehydic forms of oleuropein and ligstroside aglycon [28]. Virgin olive oil samples with elevated levels of secoiridoids exhibits greater resistance to oxidation (higher oxidative stability) and higher bitterness intensity.

2.1.3 Phenolic acids

Typically, the term "phenolic acids" refers to the phenolic compounds having one carboxylic acid group. They are mainly divided into two sub-classes: hydroxybenzoic acids (derived from benzoic acid) and hydroxycinnamic acids (derived from cinnamic acid). The most commonly phenolic acids detected in virgin olive oil are: protocatechuic, *p*- and *o*-coumaric, *p*-hydroxybenzoic, caffeic, gallic, cinnamic, vanillic, syringic, and ferulic acids [28, 29]. As compared to other chemical classes, phenolic acids are generally found in lower concentrations in virgin olive oil. Nevertheless, these compounds are acknowledged as strong natural antioxidants displaying a significant role in wide range of biological properties and sensory features of virgin olive oil [29].

2.1.4 Flavonoids

Virgin olive oil flavonoids compounds are widely appreciated for their beneficial health-related effects. They possess two aromatic rings linked by a linear three carbon chain. The main identified flavonoids in this product are luteolin, apigenin and also, even at very low concentration, methoxyluteolin [24].

2.1.5 Lignans

The chemical structure of these metabolites is generally formed by the combination of two units of phenylpropane (carbon 6- carbon 3). (+)-1-acetoxypinoresinol and (+)-1-pinoresinol are the main lignans found in virgin olive oil [30].

2.1.6 Hydroxy-Isochromans

They are present at low concentrations, being mainly formed during the malaxation step through the interaction of hydroxytyrosol and aromatic aldehydes [29, 31]. Mainly two hydroxy-Isochroman compounds were detected in virgin olive oil: 1-phenyl-6,7-dihydroxyisochroman and 1-(3′-methoxy-4′-hydroxy)phenyl-6,7-dihydroxyisochroman [31].

2.2 Major factors affecting virgin olive oil phenolic fraction

Recent attempts to boost virgin olive oil consumption are focusing on the promotion of large range of its quality traits with a specific focus on sensory features and health value. Interestingly, there is huge interest, among producers, in improving organoleptic and nutritional characteristics to meet consumer's demand for healthy premium quality virgin olive oil. These features are acknowledged as the result of complex interactions between biotic and abiotic factors that regulate the biosynthesis and the amounts of key aroma and bioactive compounds in this matrix. Following that logic, seems to make sense that a first steps towards the development of holistic promising approach to produce high organoleptic quality and healthy virgin olive oil, should start with a deep investigation of the metabolites involved, focusing particularly on their qualitative and quantitative variability in response to endogenous and external factors. Doing so, it could provide crucial information for a better understanding of the mechanisms underlying the content of virgin olive oil on these compounds and, by the way, the impact of various agro-technological factors on virgin olive oil sensory quality and health value.

In this line, for their presumed roles in determining the main virgin olive oil nutritional characteristics along with their implication on enhancing the oxidative stability of this product as well as their great influence on its overall sensory quality (phenolic compounds in combination with volatile constituents are mainly responsible of virgin olive oil's astringency and bitterness), great attention has been paid to understanding those factors responsible of their qualitative and quantitative variations. Several works have been carried out in this respect with a final goal to modulate the impact of these factors to promote the production of a virgin olive oil with high phenolics content. Undoubtedly these compounds are among the most investigated virgin olive oil constituents during the last two decades.

As mentioned above, precursors of virgin olive oil phenolic compounds are biosynthesised during fruit development, resulting from many interacting fruit's growth processes and metabolic and enzymatic activities, which are regulated by internal and external factors such as cultivar, edaphoclimatic growing conditions, cultivation techniques, and ripening process. Once harvested, post-harvest processing (crushing, malaxation, centrifugation, filtration, and storage conditions) factors induce the production and/or the loss of different types of phenolic compounds.

2.2.1 Genetic factors

A large diversity of olive cultivars is used worldwide to produce virgin olive oils with distinct characteristics. Genotype has been pointed as one of the most important factors which significantly influence the phenolic composition of virgin olive oil [32]. As such, the effect of cultivar on this fraction has been attributed to differences in genetic characteristics. Phenolic profiles of virgin olive oil from different varieties and countries obtained by using different agronomic and technological tools have been extensively studied in the literature. In a recent three years study realised by Miho et al. [33] on mono-varietal virgin olive oil obtained from 44 cultivars, the results showed a great qualitative variability in the phenolic composition among the studied cultivars. The genotype was responsible of the highest proportion of variance (66.79%) while the inter-annual parameter explained only 3.67% of variance. Besides, although it is generally assumed that qualitatively, the phenolic composition of virgin olive oils is the same regardless the variety used, however, the existence of quantitative variation of phenolics among olive cultivars has been cited [22, 23]. These differences are outlined as critical information when dealing with the prediction of virgin olive oil stability against oxidation and its organoleptic features. In this sense, cultivars with high phenolic content are expected to be more bitter taste (high intensity of bitterness and pungency) and show higher shelf life and health value [34]. In this sense, Beltran and co-workers [35] suggest a classification scale of virgin olive oils based on their total phenols content expressed as milligramme of caffeic acid per kilogramme of oil. Accordingly, virgin olive oil with phenols level lower than 220 mg/kg is considered as non-bitter oil or with imperceptible bitterness; a slight bitter oil shows a phenolics content ranging from 220 to 340 mg/kg; bitter oils are characterised by a total phenols content varying between 340 and 410 mg/kg; whereas in bitter or very bitter virgin olive oil, richness on phenolics exceeds 410 mg/kg.

Table 1 contains data on the total phenolic contents and antioxidant activities of virgin olive oils obtained from the major olive cultivars growing in the main

Country	Cultivar	Total phenolic content	ABTS assay	DPPH assay	FRAP assay	References
Spain	Picual	419-671[a]	3.46[c]	0.40[c]	1.15[c]	[36, 37]
	Cornicabra	317[a]	2.62[c]	0.60[c]	0.94[c]	[36]
	Arbequina	104-302[a]	0.2–2.09[c]	0.44–1.58[c]	0.54–2.22[c]	[36, 38, 39]
Italy	Coratina	112-532[b]	21.37[d]	35.82[d]	82.32[d]	[40, 41]
			0.27–0.57[e]	NA	0.32–0.63[e]	
	Frantoio	94.6-256[b]	17.9–36.4[i]	56.5 - 106[d]	91.3 – 156[f]	[42, 43]
Greece	Koroneiki	116–373.3[b]	0.25–0.44[e]	NA	0.30–0.54[e]	[41, 44]
Tunisia	Chemlali	4.3–11.52[h]	0.7–3.03[i]	2.69–25.87[d]	NA	[45, 46]
		158[b]	0.6[c]	37.23[g]		
	Chétoui	3.46–9.24[h]	0.25–2.2[i]	10.23–184.12[d]	NA	[45, 46]
		395[b]	2.42[c]	78,56[g]		
Portugal	Galega	118-137[b]	0.33[e]	NA	0.42[e]	[41, 47]
			0.35[c]	NA	NA	
Turkey	Memecik	296.24–407.13[a]	1.18–1.86[c]	0.50–0.81[c]	NA	[48, 49]
		95.86[e]				
	Gemlik	150.92–245.40[a]	0.76–1.24[c]	0.32–0.44[c]	NA	[48, 49]
		42.17[e]				
	Ayvalik	93.19[e]	1.34–2.26[c]	0.53–1.05[c]	NA	[41, 48]
		91-130[b]	0.16–0.28[e]	NA	0.2–0.3[e]	
Morocco	Picholine Marocaine	216.83–668.67[a]	NA	NA	NA	[50, 51]
		112-390[b]				

Results are expressed as: [a]: mg caffeic acid/kg oil; [b]: mg gallic acid/kg oil; [c]: mmol de Trolox /kg oil; [d]: IC 50 µg/mL; [e]: mg/kg oil; [f]: µM Fe (II)/g; [g]: percentage; [h]: g/kg of fresh olive fruit; [i]: mmol/L TEAC; NA: not analysed/not reported. Abbreviations: ABTS: 2,2-azino-bis(3-ethylbenzothiazoline-6-sulfonic acid; DPPH: 2,2-diphenyl-1-picrylhydrazyl; FRAP: ferric reducing ability of plasma.

Table 1.
Total phenolic contents and antioxidant properties of virgin olive oil.

virgin olive oil producing countries. Data is taken from selected scientific papers published between 2010 and 2020. To quantify total phenolic contents, all researchers use Folin–Ciocalteu reagent and except a few report, total phenolic contents are reported as gallic acid equivalents (GAE)/kg.

As can be seen from this Table, the levels of phenolics are highly variable. "Picual", "Koroneiki" and "Picholine Marocaine" are among the main olive varieties with the highest phenolic content compared to the lowest levels recorded in "Arbequina" virgin olive oil. It has been hypothesised that the differences in phenolic content between olive varieties may be attributed to the differences in the expression of genes encoding for enzymes involved in these metabolites biosynthesis's pathway during olive growth and ripening, and in the activity of these enzymes during oil processing. Thereby, some monovarietal virgin olive oils present peculiar and specific taste induced by their phenolic fraction deriving from olive fruit or generated during processing. As such, phenolic profiles start to be used as effective varietal markers to classify virgin olive oils according to their varietal origin [52, 53]. Moreover, in the current globalised and ever more fiercely competitive global virgin olive oil market, competitiveness can mainly be maintained by investing in differentiation strategies. In this regard, producing monovarietal virgin olive oil with high phenolics content and specific sensory properties is a common differentiation strategy used worldwide. These peculiar characteristics of this kind of virgin olive oil are the major driving forces for their high economic value. However, high priced monovarietal virgin olive oils are frequently subjected to fraud and mislabelling practices, therefore, recently there has been a growing interest in developing analytical approaches to determine the varietal origin of virgin olive oil. To this end, varietal authentication based on phenolic profiling and/or fingerprinting is one of the reliable tools used nowadays [52, 53].

2.2.2 Geographical origin

The term "geographical origin" must be taken to include both grown region edaphoclimatic conditions and olive growing agronomic practices. In this respect, it has been elucidated that virgin olive oils obtained from the same cultivar, growing in different geographical locations, show distinctive phenolic profiles. Consequently, various phenolic profiling and fingerprinting approaches were successfully applied to authenticate the geographical origin of this product [54, 55].

Environmental factors may influence widely the phenolic profile of olive fruits before harvesting. Indeed, when it undergoes stressful environmental conditions, olive tree responds by increasing the biosynthesis of antioxidant secondary metabolites such as phenolic compounds to combat oxidative damage induced by stress. It is also proposed that abiotic stress induces profound changes in the expression of endogenous olive fruit enzymes activities which in turn will affect the amount phenolics in the obtained oils. Thus, when dealing with the impact of environmental factors on virgin olive oil phenolic fraction, studies have mainly focused on the effect of edaphoclimatic conditions such as soil type and salinity, altitude, and climatic conditions. Among these factors, water availability has been gaining prominence in the literature due to its influence on olive fruit growth and development, especially in Mediterranean regions where rainfalls and water resources are limited. In this regard, there is ample scientific evidence about the noticeable increase of virgin olive oil total phenols content under drought stress. In this sense, Romero and Motilva [56] reported high total phenols content in oils extracted from olive fruits that have undergone a summer drought stress. Similarly, high phenolics concentration could be expected in virgin olive oil produced in semi-arid and arid region characterised by extremely hot climates during the summer period [57].

Furthermore, cultivation conditions have been shown to influence greatly the levels of virgin olive oil phenolic compounds. Water deficit generates a stress situation that induces the production of phenolics. For instance, by applying a water deficit irrigation strategy (46 to 48% of water requirements calculated from reference evapotranspiration using a crop coefficient of 0.55), Caruso and co-workers [58] reported higher contents of total phenolics and secoiridoids derivatives when compared to oils obtained from fruits of fully irrigated trees [58]. These findings explain why virgin olive oils obtained from fully irrigated trees tend to be less bitter and pungent than those obtained from rainfed olive orchards if bearing in mind the strong correlation between these organoleptic features and phenolics amounts in virgin olive oil.

In addition, by investigating the effect of geographic origin latitude on the phenolic content in Tunisian virgin olive oil, Issaoui and co-authors [59] reported that oils obtained from olive cultivars growing in high altitudes had three times higher phenolic content than oils from the same cultivars grown at low altitudes.

Fertilisation is another cultural practice affecting the biosynthesis of phenolics in olive fruits. The effect of fertiliser application on virgin olive oil's phenolic fraction has been cited, being strongly depending on the application period, and nutrients type and doses. For instance, previous scientific findings reported a significant decrease in the total phenolics content in virgin olive oil with an increased amount of nitrogen fertilisation supplied even by foliar application during specific fruit development stages or *via* irrigation water throughout the crop season [60, 61]. Some authors suggest a protein-phenol competition to explain this significant decrease of total phenols induced by nitrogen fertilisation excess [61]. Hence the importance of a balanced controlled nitrogen fertilisation to obtain virgin olive oils with a high total phenolics content.

2.2.3 Harvest time

Besides varietal and geographic origin dependent variations, significant differences in virgin olive oil phenolics content have been observed according to the corresponding olive fruit ripening stage at harvest. In this context, it is important to remember that the predominant phenolic compounds detected in virgin olive oil, i.e. the secoiridoids derivatives, result from the enzymatic hydrolysis of the glycosylated form of oleuropein, demethyloleuropein and ligstroside compounds naturally present in olive fruits. Overall, it can be concluded that the phenolic amount of the olive fruits considerably affects the concentrations and proportions of various phenols of the oil that will be obtained after mechanical extraction [62]. Therefore, olive harvesting at the most appropriate stage of ripening is crucial to maximise the phenolic content of virgin olive oil. Extensive research has been conducted to investigate the evolution of phenolics content and composition in olive fruits and corresponding oils during ripeness [63]. Although many of these previous studies have shown that phenolics concentration decrease sharply during ripening (oils obtained from olive fruits harvested at earlier maturity stages show the highest contents of phenolic antioxidants when compared with those derived from fully ripe olives), contrasting data to such a pattern have been also cited [64]. In this sense, several authors found that the influence of olive ripening stage on virgin olive oil phenolics content is cultivar dependent [65].

2.2.4 Oil processing

The phenolic profile of virgin olive oil depends not only on olive fruits composition at harvest but also on the changes occurring during oil processing. Virgin olive oil extraction mainly involves preliminary operations (fruits reception and cleaning); paste preparation by means of breaking the fruit structure (crushing); the liberation of the oil from the cells and the formation of solid and liquid phases (malaxation); separation of the solid (pomace) and liquid phases (oil and/or wastewater) by pressing or horizontal centrifugation, and separation of the liquid phases (oil and wastewater) by decantation or vertical centrifugation. There are many reports in the literature related to the impact of each one of these steps on virgin olive oil phenolic profile suggesting that the latter is strongly determined by some processing conditions such as crushing method (stone mill or hammer or disk crusher) and conditions (grid holes diameter and rotation speed), the malaxation time and temperature and the presence or absence of oxygen during malaxation. Recent and exhaustive reviews are available in literature for getting more comprehensive detailed information about the impact of processing conditions on virgin olive oil phenolic compounds [66, 67]. It is worth underlining that some of these variables have been studied with the aim of controlling and/or modulating enzymatic process affecting the phenolic fraction, mainly the action of the β-glucosidase, phenoloxidases and esterases, which hydrolyse precursors of oleuropein and ligstroside to produce secoiridoids derivatives, the main phenolic group in virgin olive oil, and therefore the phenolic concentration and the sensory characteristics associated with them. For instance, Antonini et al. [68] revealed that centrifugation by means of a two-phases decanter system produced oils with higher concentrations of oleacein, oleocanthal, oleuropein aglycone, lignans, (+)-pinoresinol and (+)-1-acetoxypinoresinol when compared with three-phases decanters. Similarly, the contents of oleacein and oleocanthal raised linearly with crushing speed [69]. Regarding malaxation conditions, shorter malaxation time (30 to 45 min) was widely associated with higher contents of phenolic compounds regardless of the cultivar [70]. In contrast, rising the malaxation temperature from 27 to 47°C induced higher concentrations of hydroxytyrosol, tyrosol, pinoresinol and *p*-coumaric acid [71].

2.2.5 Oil storage conditions

Once virgin olive oil is extracted, storage conditions are the next critical points to be considered for preserving its phenolic fraction. In fact, virgin olive oil shelf life is quite related to its phenolic content being exposed to a significant decline if this product is stored under inappropriate conditions. Lolis and co-workers [72] recently approved the fact that dark glass containers and low temperature would maintain virgin olive oil phenolic composition for up to 9 months while a temperature of 37°C was associated to a rapid deterioration of its quality after only 3 months. The same trend was reported by Li et al. [73] who revealed that cold storage conditions induced a better preservation of hydroxytyrosol, tyrosol and oleuropein contents when compared with optimal temperature (25°C).

To conclude this section, it should be underlined that, even though virgin olive oil phenolic fraction is very complex due to its origin and the large number of factors that influence its amount in this product, it remains one of the most determining factors of health value and organoleptic quality of virgin olive oil. Thereby, understanding the quantitative and qualitative variability of this fraction is of paramount importance when dealing with putting in place an agro-technological strategy to produce a virgin olive oil which phenolic content fulfils the EFSA health claim requirements. Nevertheless, accurate measurement of virgin olive oil phenolics content remains a challenging task since no official analytical method has been yet considered by any regulatory body. To fill this gap, tremendous effort has been put during the last decades and a great variety of analytical methodologies have been developed and implemented so far. Most common methods and recent advancements are succinctly described in the following section. An in-depth discussion of analytical methodologies applied to virgin olive oil phenolic fraction characterisation is beyond the scope of this chapter. We suggest the reading of some interesting reviews book chapters and research articles previously published, where the authors give an exhaustive overview on this topic [9, 74, 75].

2.3 Advancement in analytical techniques for the characterisation of virgin olive oil phenolic fraction

The need for highly sensitive and selective analytical methods for the determination of phenolics and checking their content in commercialised virgin olive oils has emerged the development of several analytical methodologies from targeted analysis to metabolomics-based approaches. Thus, substantial developments in research focused on the extraction, separation, identification, and quantification of virgin olive oil phenolic compounds have occurred over the last decades. To date, there is no universal workflow described for all analytical approaches dealing with the characterisation of phenolics in virgin olive oil. However, regardless of the applied approach, a typical analytical method consists of several steps such as sampling, sample preparation, separation, detection, and data analysis.

A carefully samples preparation and extraction of phenolics from virgin olive oil samples is a critical step to avoid the loss of these analytes which can lead to significant errors. Before the extraction and isolation of these components, samples must be collected, preserved, and properly prepared. Commonly, phenolic compounds extraction prior to their identification and quantification is necessary. Currently, the main extraction methods used to this end are liquid–liquid extraction (LLE) and solid-phase extraction (SPE). Even though the first technique is an easy-operating method, it remains quite inefficient and time-consuming. In contrast, SPE is a less solvent-consuming and effective method but considered as labor-intensive. Selecting the most pertinent method depends on the scope of the analysis, the expected concentrations, and the analytical platform to be used.

The recent progress in the so-called "Metabolomics" and the continuous effort towards the development of new high-throughput technologies induced a great advancement in analytical techniques developed for the characterisation of virgin olive oil phenolic fraction. Thus, although the traditional colorimetric assay (also known as Folin–Ciocalteu assay) remains the most widely used technique for the quantitative determination of the total phenolics content in this matrix, a deep characterisation of its phenolic profile requires a prior separation, usually done by reversed phase LC coupled to diode array detector (DAD). For identification purposes, the use of MS is becoming increasingly popular [74]. Other alternatives have also emerged such as ultra-high-performance liquid chromatography (UHPLC) and gaz chromatography (GC) coupled to MS (GC–MS), nuclear magnetic resonance (NMR) and other spectroscopic techniques [75, 76].

Folin–Ciocalteu method is mainly based on the interaction between the functional hydroxyl groups of phenolic compounds and the Folin–Ciocalteu reagent, which consists of a mixture of phosphortungstate and phosphomolybdate. This colorimetric method is simple and offers a great reproducibility and repeatability. However, low specificity is considered as its main drawback [77]. Din et al. [78] recently developed a robust methodology combining SPE and the traditional Folin–Ciocalteu method for the quantification of total phenolic content in virgin olive oil. Phenolic compounds recovery was higher than 95% and the quantification through the colorimetric assay was linear, repeatable, reproducible and precise in 100–500 ppm measuring range.

Otherwise, to overcome these drawbacks, a variety of analytical methodologies have been proposed and successfully applied taking advantage of recent developments in LC and GC instrumentation coupled to different detection systems which in turn allowed increasing the selectivity and sensitivity of virgin olive oil phenolics analysis. In this sense, various MS-based strategies, NMR and vibrational spectroscopy approaches have emerged. NMR spectroscopy is continuously gaining interests as a powerful and robust analytical platform. It offers the ability for a rapid and high screening of the phenolic profile leading to a better detection of all possible changes occurring in response to various factors such as storage, packaging, environmental, agronomic, genetic and processing conditions, etc. Interesting examples to illustrate the application of NMR spectroscopy in virgin olive oil phenolics analysis were reported in a review paper by Dais et al. [79]. Nevertheless, despite its high accuracy, non-destructive nature and limited sample preparation needs, NMR remains an underused technique in virgin olive oil analysis due to its low sensitivity and the need for highly skilled scientists for spectral interpretation [80].

MS based metabolomic approaches (using LC–MS, GC–MS or capillary electrophoresis (CE-MS)) have been largely applied for the quantitative and qualitative analysis of virgin olive oil phenolic fraction [74, 76]. CE offers high resolution, fast analysis speed and relatively low operating cost [81]. However, poor sensitivity and low reproducibility are its main disadvantages, precluding its use for the analysis of phenolic compounds at very low concentrations [82]. In contrast, GC is believed to be one of the most sensitive and reproducible separation technique used in virgin olive oil analysis [76]. GC–MS approaches are widely used for the accurate analysis of volatile and non-volatile organic metabolites. Yet, the need for a derivatisation step is mandatory for non-volatile compounds, including phenolic compounds, which is obviously the major drawback of this analytical platform [76]. This fact has boosted the extensive use of other alternative separation methods mainly LC. This technique is considered as "the workhorse" tool for phenolic compounds separation due to their non-volatile character. Researchers in the field of virgin olive oil analysis seems to steadily adopt LC methodologies owing to their simplicity, rapidity, and high sensitivity [76]. Combined with highly sensitive and selective MS detection, it could provide the simultaneous separation, identification and quantification of different phenolic compounds occurring in virgin olive oil. Thus, various LC–MS phenolic profiling or fingerprinting approaches were developed and successfully applied for quality control and authentication purposes as well as to assess the bioavailability of this product [74].

3. Virgin olive oil phenolic compounds: health-promoting effects and bioavailability

Phenolics are believed to confer a wide range of benefits to virgin olive oil (**Figure 2**) due to their powerful antioxidant activity (**Table 1**). Thus, Since both oxidative stress and inflammation are considered as major contributing factors to neurodegenerative and cardiovascular diseases, virgin olive oil phenolics are thought to exhibit a strong anti-inflammatory effect and directly participate to the redox balance of human cells [28]. For instance, hydroxytryrosol, oleuropein and their derivatives may exert a protective effect against the amyloid plaque generation [83]. Similarly, oleuropein aglycone and oleocanthal have been shown to interact with Aβ aggregation states leading to a better protection against Alzheimer disease [84]. As revealed in Predimed trials, a regular consumption of virgin olive oil induces a significant decrease in risk parameters of developing cardiovascular diseases, namely inflammatory cytokine, the vascular cell adhesion molecule and intercellular adhesion molecule, as well as a rise in high-density lipoprotein (HDL) levels and reduced LDL levels [85]. In particular, the phenolic compounds present in this vegetable oil are related to the inhibition of lipid peroxidation induced by free radicals [86]. It has also been noted that hydroxytyrosol can improve the levels of circulating lipids and repairing oxidative damage which is responsible for numerous cardiovascular issues [87].

Consumption of virgin olive oil with a high content on phenolic antioxidants may also be effective in cancer prevention. Oleuropein glucosides, hydroxytyrosol, and to a lesser degree tyrosol, were highlighted in an interesting review paper by Casaburi et al. [88] as the most promising chemopreventive agents among olive oil phenols against cancer, mainly for their role in preventing DNA oxidative damage in various human cell types. The same researchers revealed the fact that olive oil phenolic compounds induce anti-tumour effects against various tumour types including leukaemia, colorectal and breast cancer due to their ability to inhibit proliferation and enhance apoptosis.

Another potential therapeutic effect of virgin olive oil phenolics includes the prevention and treatment of type 2 diabetes. This was explained by the capability of these compounds to decrease insulin resistance *via* the inhibition or reduction of

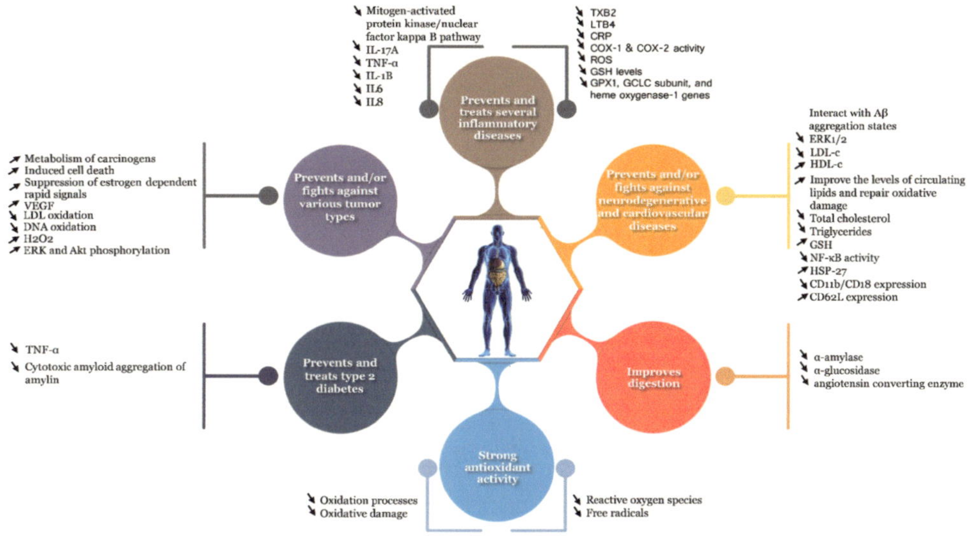

Figure 2.
Main mechanisms involved in the health properties of virgin olive oil phenolic compounds.

pro-inflammatory molecules such as TNF-α [89]. Besides, hydroxytyrosol and oleuropein have been shown to exhibit an antimicrobial effect against several bacterial strains. Indeed, although they were more effective against ATCC bacterials trains including *Hemophilus influenzae* ATCC 9006, *Moraxella catarrhalis* ATTC 8176, *Salmonella typhi* ATCC 6539 and *Staphylococcus aureus* ATTC 25923, these molecules induced a promising cytotoxic activity against a great number of bacterial strains such as *Salmonella* spp., *Vibrio alginolyticus and Vibrio cholerae* [90].

Nevertheless, even virgin olive oil functional value is strongly determined by its phenolic content, however, only a percentage of this content can be biologically active in the body, as it must be absorbed through the gastrointestinal tract and reach the bloodstream, which is known as bioavailability. The latter may simply be defined as the fraction that reach the organism where it can pursue its biological effect, may be enhanced or inhibited depending on various factors such as chemical structure, concentration, interaction with other compounds, enzymes activity, host's age and physical condition, etc. [91].

However, in the case of virgin olive oil most investigations have focused on some specific phenolic compounds such as oleuropein, tyrosol, hydroxytyrosol, and their derivatives as they were associated with the highest bioactivity. Outcomes of selective recent studies of the bioavailability of these compounds are presented in **Table 2**.

Thus, after intake, virgin olive oil forms a micellar solution in the gastrointestinal tract. Most olive oil phenolic compounds pass through the mouth and stomach to reach the small intestine and colon without any modification [89]. Hydroxytyrosol and tyrosol have been demonstrated to be the most absorbed phenolics in the intestinal tract (absorption rate is 40 to 95% approximately) [107]. Also, Corona et al. [108] suggested that the amounts of hydroxytyrosol and tyrosol that reach the small intestine following incubation and passage through the acidic conditions of the stomach are considerably higher than those initially present in ingested virgin olive oil. The recovery of these two phenolic compounds is also widely related to the biological matrix in which they were administrated. Indeed, urinary recovery of hydroxytyrosol was greater after virgin olive oil intake than after the addition of hydroxytyrosol to a yoghurt or to a refined olive oil [109]. The urinary hydroxytyrosol and tyrosol are absorbed from a moderate and sustained dose of virgin olive oil

Phenolic compound	Health effect	Dose	Cell type and/or animal model.	Mechanism	Reference
Hydroxy-tyrosol	Protection against breast, prostate and colon cancers.	100 µM.	Breast: MDA and MCF-7; Prostate: LNCap and PC3; Colon: SW480 and HCT116 cancer cells.	↑ H2O2.	[92]
	Prevention of neurodegenerative diseases.	50 µM.	N2a neuroblastoma cells.	↓ NF-κB activity. ↑ GSH (at 100 µM).	[93]
	Prevention and treatment of metabolic syndrome.	10 and 50 mg/kg/day.	C57BL/6J mice.	↓ SREBP-1c/FAS pathway. ↑ Antioxidant enzyme activities. Normalise expression of mitochondrial complex subunits and mitochondrial fission marker Drp1.	[94]
	Prevention of cardiovascular diseases.	0.1–10 µg/ml.	H9c2 cells.	↓Phosphorylation of a transcriptional target c-Jun. ↑ Phosphorylation of extracellular signal-regulated kinase 1/2. ↑ Heat shock proteins (HSP)-27.	[95]
Tyrosol	Prevention of neurodegenerative diseases.	50 µM.	N2a neuroblastoma cells.	↓ NF-κB activity.	[93]
	Anti-inflammatory effect.	100 µM (*in vitro* study) and 0.1–0.5 mg/kg (*in vivo* study).	Human umbilical vein endothelial cells, a human monocytic cell line THP-1 and male Swiss albino mice.	↓ ROS. ↓ GSH levels. ↓ GPX1, GCLC subunit, and heme oxygenase-1 genes.	[96]
Oleuropein	Prevention of cardiovascular diseases.	100 µM.	Vascular smooth muscle cells.	↓ ERK1/2.	[97]
	Anti-cancer effect.	125 mg of oleuropein / kg of diet.	MCF-7 cells xenograft and female nu/nu athymic 7–8 week old mice.	Unknown.	[98]
	Prevention of neurodegenerative diseases.	Unknown	—	↓ Senile plaque formation through amyloid beta peptide (Aβ) aggregation.	[99]
	Treatment of skin diseases and wounds.	50 mg of oleuropein / kg/day.	Male Balb/c mice.	↑ VEGF.	[100]

Phenolic compound	Health effect	Dose	Cell type and/or animal model.	Mechanism	Reference
Oleacein	Anti-cancer effect.	1–100 μM.	Human epidermoid carcinoma cell line A431 and human immortalised keratinocytes.	↓ Erk and Akt phosphorylation. Suppression of B-Raf expression.	[101]
	Anti-inflammatory effect.	50–100 μM.	Human monocytes.	↓ Cox-2. ↓ Superoxide anions production.	[102]
	Prevention of cardiovascular diseases.	50–100 μM.	Human isolated neutrophils.	↓ Neutral endopeptidase activity, elastase, metalloproteinase 9 and IL 8 (100 μM). ↓ CD11b/CD18 expression (50 μM). ↑ CD62L expression (50 μM).	[103]
Oleocanthal	Prevention of neurodegenerative diseases.	Ranging between 0.1 and 25 μM.	—	Inducing stable conformational modifications of tau-441 protein secondary structure and interfering with tau aggregation.	[104]
	Anti-cancer effect.	Between 5 and 80 μM for the *in vitro* study and up to 10 mg/kg for the *in vivo* study.	Human hepatocellular carcinoma cells and Male BALB/c athymic nude mice.	↓ Epithelial-mesenchymal transition. ↓ Transcription factor STAT3 nuclear translocation. ↓ DNA binding activity.	[105]
		1–100 μM.	Human epidermoid carcinoma cell line A431 and human immortalised keratinocytes.	↓ Erk and Akt phosphorylation. Suppression of B-Raf expression.	[101]
	Anti-inflammatory effect.	50 μM.	ATDC5 murine chondrogenic cells and murine macrophages J774.	↓ MIP-1α inflammatory mediator at the protein and mRNA level. ↓ IL-6 inflammatory mediator at the protein and mRNA level.	[106]

Abbreviations. Cox-2: cyclooxygenase2; ERK1/2: extracellular signal-regulated kinase 1/2; GCLC: glutamate-cysteine ligase catalytic; GPX1: glutathione peroxidase 1; GSH: glutathione; IL 8: interleukin 8; NF-κB: nuclear factor kappa-light-chain-enhancer of activated B cells; ROS: reactive oxygen species; VEGF: vascular endothelial growth factor; ↓: decreasing trend; ↑: increasing trend.

Table 2.
Selective studies on bioactivities and potential health benefits of main virgin olive oil phenolic compounds.

which is similar to that consumed daily in a typical Mediterranean diet [110]. These simple phenols were largely identified in both urine and plasma mainly in glucuronides and sulphate conjugates, while their free forms were not detected in plasma samples. In fact, aglycones and glycosides forms of tyrosol and hydroxytyrosol undergo a prompt hydrolysis phenomenon under gastric conditions together with a substantial rise in the contents of tyrosol and hydroxytyrosol free forms penetrating the small intestine [108]. The intestinal transport of hydroxytyrosol occurs through a bidirectional passive diffusion mechanism as demonstrated in an *in vitro* study conducted by Manna and co-workers [111]. While crossing epithelial cells of the gastrointestinal tract, hydroxytyrosol is usually transformed through enzymatic reactions into homovanillyl alcohol and its glucuronide forms [108].

Regarding secoiridoids, they remain highly stable in the mouth but suffer significant losses in the gastric, duodenal, and colonic regions, with a recovery rate at the duodenal level ranging between 7% and 34%. Glycosylation and cleavage of glycosidic linkages take part in the secoiridoids absorption, and it is thought that some of them, such as oleacein, are absorbed in the small intestine by passive diffusion through the membrane of intestinal cells [89]. In the case of oleuropein, the mechanism of absorption is still confusing and remains unclear. However, several studies showed that it exerts its biological effects *via* its conversion into hydroxytyrosol [112]. This was mainly explained by the fact that the content of oleuropein was in the mass range of few nanograms in plasma while hydroxytyrosol was detected in high concentrations after the intake of great doses of oleuropein in both rats and humans trials [113, 114]. When considering other secoiridoids, Vissers et al. [114] could not analyse oleuropein glycoside, oleuropein- or ligstroside-aglycons in urinary excretions which supports the idea that they may be hydrolysed into hydroxytyrosol and tyrosol and extensively metabolised once absorbed from the small intestine.

4. Concluding remarks

In the Mediterranean area, the largest virgin olive oil producing region in the world, new agronomic practices and processing technologies are steadily developed and adopted over the past decade. Thereby, virgin olive oil production has achieved outstanding performance both in terms of increasing oil yield and quality. Nevertheless, the future of this sector in a context of globalisation, and the consequent changes in lifestyle and consumers adherence to the Mediterranean diet, has become an important subject of attention for the producers and governmental bodies alike. For instance, in the search for new opportunities to boost virgin olive oil consumption, growing importance is attributed to promote its health value and organoleptic features among consumers, paying special attention to those compounds responsible of these characteristics. In developing such strategies, a central role should clearly be reserved for the phenolic fraction of this product if considering its health-promoting proprieties and its contribution to the oxidative stability and sensory quality of virgin olive oil. Undoubtedly, the recent agro-technological advancements supported by the scientific data available up till now offer promising tools to produce virgin olive oils rich on phenolics; however, the bitter taste of such oils is often not appreciated by consumers. For this reason, what still remains a challenge in this sector is the development of holistic agro-technological approaches to produce virgin olive oils with phenolics content that comply not only with institutional regulations (EFSA health claim for example) but also with consumer preferences. Furthermore, the lack of an official method of determining these compounds in compliance with health claim requirements is an obstacle to be overcome in a not-too-distant future if we want to avoid a lack of credibility of these health claims.

Acknowledgements

Authors are grateful to the Erasmus+ Programme of the European Union, FoSaMed project (618518-EPP-T-2020-1-PT-EPPKA2-CBHE-JP) for Salah Chaji's predoctoral fellowship; and to the National Centre for Scientific and Technical Research (CNRST Morocco) for Sanae Moussafir's predoctoral fellowship (Research Excellence Fellowships Program).

Author details

El Amine Ajal[1], Salah Chaji[2], Sanae Moussafir[2], Rachid Nejjari[1], Abdelmajid Soulaymani[3] and Aadil Bajoub[2,3]*

1 UPR of Pharmacognosy, Faculty of Medicine and Pharmacy of Rabat, Mohammed V University, Rabat, Morocco

2 Laboratory of Food and Food By-Products Chemistry and Processing Technology, National School of Agriculture in Meknès, Meknès, Morocco

3 Laboratory of Health and Biology, Faculty of Sciences, University Ibn Tofail, Kénitra, Morocco

*Address all correspondence to: aliam80@hotmail.com

References

[1] Farràs M, Martinez-Gili L, Portune K, Arranz S, Frost G, Tondo M, et al. Modulation of the gut microbiota by olive oil phenolic compounds: Implications for lipid metabolism, immune system, and obesity. Nutrients. 2020;12(8):2200. DOI: 10.3390/nu12082200

[2] Fitó M, Cladellas M, de la Torre R, Martí J, Muñoz D, Schröder H, et al. Anti-inflammatory effect of virgin olive oil in stable coronary disease patients: A randomized, crossover, controlled trial. Eur J Clin Nutr. 2008;62(4):570-574.

[3] Buckland G, Travier N, Barricarte A, Ardanaz E, Moreno-Iribas C, Sánchez MJ, et al. Olive oil intake and CHD in the European Prospective Investigation into Cancer and Nutrition Spanish cohort. Br J Nutr. 2012;108(11):2075-2082. DOI: 10.1017/S000711451200298X

[4] Guasch-Ferré M, Hu FB, Martínez-González MA, Fitó M, Bulló M, Estruch R, et al. Olive oil intake and risk of cardiovascular disease and mortality in the PREDIMED Study. BMC Med. 2014;12(1):78. DOI: 10.1186/1741-7015-12-78

[5] Covas MI, Ruiz-Gutiérrez V, De La Torre R, Kafatos A, Lamuela-Raventós RM, Osada J, et al. Minor components of olive oil: Evidence to date of health benefits in humans. Nutr Rev. 2006;64(10):S20–S30.

[6] Preedy VR, Watson RR, editors. Olives and olive oil in health and disease prevention. 2nd edition. Elsevier; 2021.

[7] Boskou D, editor. Olive and olive oil bioactive constituents. 1st edition. American Oil Chemists' Society Press; 2015.

[8] Jimenez-Lopez C, Carpena M, Lourenço-Lopes C, Gallardo-Gomez M, Lorenzo JM, Barba FJ, et al. Bioactive compounds and quality of extra virgin olive oil. Foods. 2020;9(8):1014. DOI: 10.3390/foods9081014

[9] Rodríguez-López P, Lozano-Sanchez J, Borrás-Linares I, Emanuelli T, Menéndez JA, Segura-Carretero A. Structure – biological activity relationships of extra-virgin olive oil phenolic compounds: Health properties and bioavailability. Antioxidants. 2020;9(8):685. DOI: 10.3390/antiox9080685

[10] Servili M, Sordini B, Esposto S, Urbani S, Veneziani G, Maio I Di, et al. Biological activities of phenolic compounds of extra virgin olive oil. Antioxidants. 2014;3(1):1-23. DOI: 10.3390/antiox3010001

[11] De La Torre R. Bioavailability of olive oil phenolic compounds in humans. Inflammopharmacology. 2008;16(5):245-247.

[12] De La Torre R, Fitó M, Covas M-I. The bioavailability of olive oil phenolic compounds and their bioactive effects in humans. In: Preedy VR, Watson RR, editors. Olives and Olive Oil in Health and Disease Prevention (second edition). Elsevier; 2021. p.193-203.

[13] Uceda M. Composición y calidad de los aceites de oliva vírgenes extra. Influencia de la variedad. In: Hernández J.V. et al., editors. Algunas contribuciones sobre olivicultura y Elaiotecnia desde la perspectiva de la experiencia. GEA Westfalia Separator Ibérica, S.A. 2009.

[14] Rodis PS, Karathanos VT, Mantzavinou A. Partitioning of olive oil antioxidants between oil and water phases. J Agric Food Chem. 2002;50:596-601.

[15] Emma MR, Augello G, Di Stefano V, Azzolina A, Giannitrapani L, Montalto G, et al. Potential uses of olive oil secoiridoids for the prevention and treatment of cancer: A narrative review of preclinical studies. Int J Mol Sci. 2021;22(3):1234.

[16] Termentzi A, Halabalaki M, Skaltsounis LA. From drupes to olive oil: An exploration of olive key Metabolites. In: Boskou D, editor. Olives and olive oil Bioactive Constituents. American Oil Chemists' Society Press; 2015. p.147-170.

[17] Fregapane G, Salvador MD. Production of superior quality extra virgin olive oil modulating the content and profile of its minor components. Food Res Int. 2013;54(2):1907-1914. DOI: 10.1016/j.foodres.2013.04.022

[18] Jerman Klen T, Golc Wondra A, Vrhovšek U, Sivilotti P, Vodopivec BM. Olive fruit phenols transfer, transformation, and partition trail during laboratory-scale olive oil processing. J Agric Food Chem. 2015;63(18):4570-4579.

[19] Velázquez-palmero D, Romero-segura C, García-rodríguez R, Hernández ML, Vaistij FE, Graham IA, et al. An oleuropein β-Glucosidase from olive fruit is involved in determining the phenolic composition of virgin olive oil. Front Plant Sci. 2017;8(November): 1902. DOI: 10.3389/fpls.2017.01902

[20] Artajo LS, Romero MP, Suarez M, Motilva MJ. Partition of phenolic compounds during the virgin olive oil industrial extraction process. Eur Food Res Technol. 2007;225:617-625.

[21] Tsimidou M. Polyphenols and quality of virgin olive oil in retrospect [*Olea europaea* L.]. Ital J Food Sci. 1998;10.

[22] Bajoub A, Medina-Rodríguez S, Olmo-García L, Ajal EA, Monasterio RP, Hanine H, et al. In-depth two-year study of phenolic profile variability among olive oils from autochthonous and mediterranean varieties in Morocco, as revealed by a LC-MS chemometric profiling approach. Int J Mol Sci. 2017;18(1):52. DOI: 10.3390/ijms18010052

[23] Miho H, Díez CM, Mena-bravo A, Medina VS De, Moral J, Melliou E, et al. Cultivar influence on variability in olive oil phenolic profiles determined through an extensive germplasm survey. Food Chem. 2018;266:192-199. DOI: 10.1016/j.foodchem.2018.06.002

[24] Di Lecce G, Piochi M, Pacetti D, Frega NG, Bartolucci E, Scortichini S, et al. Eleven monovarietal extra virgin olive oils from olives grown and processed under the same conditions: Effect of the cultivar on the chemical composition and sensory traits. Foods. 2020;9(7):904. DOI: 10.3390/foods9070904

[25] Martín-Peláez S, Covas MI, Fitó M, Kušar A, Pravst I. Health effects of olive oil polyphenols : Recent advances and possibilities for the use of health claims. Mol Nutr Food Res. 2013;57:760-771. DOI: 10.1002/mnfr.201200421

[26] Lockyer S, Rowland I. Authorised EU health claims for polyphenols in olive oil. In: Sadler MJ, editor. Foods, Nutrients and Food Ingredients with Authorised EU Health Claims. Woodhead Publishing; 2014. p.212-228.

[27] Pinto J, Paiva-Martins F, Corona G, Debnam ES, Jose Oruna-Concha M, Vauzour D, et al. Absorption and metabolism of olive oil secoiridoids in the small intestine. Br J Nutr. 2011;105(11):1607-1618. DOI: 10.1017/S000711451000526X

[28] Serreli G, Deiana M. Biological relevance of extra virgin olive oil polyphenols metabolites. Antioxidants. 2018;7(12):170. DOI: 10.3390/antiox7120170

[29] Bendini A, Cerretani L, Carrasco-Pancorbo A, Gómez-Caravaca AM, Segura-Carretero A, et al. Phenolic molecules in virgin olive oils: A survey of their sensory properties, health effects, antioxidant activity and analytical methods. An overview of the last decade. Molecules. 2007;12(8): 1679-1719.

[30] Owen RW, Mier W, Giacosa A, Hull WE, Spiegelhalder B, Bartsch H. Phenolic compounds and squalene in olive oils: The concentration and antioxidant potential of total phenols, simple phenols, secoiridoids, lignansand squalene. Food Chem Toxicol. 2000;38:647-659.

[31] Bianco A, Coccioli F, Guiso M, Marra C. The occurrence in olive oil of a new class of phenolic compounds: Hydroxy-isochromans. Food Chem. 2002;77(4):405-411. DOI: 10.1016/S0308-8146(01)00366-1

[32] Taticchi A, Esposto S, Servili M. The Basis of the Sensory Properties of Virgin Olive Oil. In: Monteleone E, Langstaff S, editors. Olive Oil Sensory Science. John Wiley & Sons; 2014. p.33-54. DOI: 10.1002/9781118332511.ch2

[33] Miho H, Moral J, Barranco D, Ledesma-Escobar CA, Priego-Capote F, Díez CM. Influence of genetic and interannual factors on the phenolic profiles of virgin olive oils. Food Chem. 2021;342:128357. DOI: 10.1016/j.foodchem.2020.128357

[34] Pedan V, Popp M, Rohn S, Nyfeler M, Bongartz A. Characterization of phenolic compounds and their contribution to sensory properties of olive oil. Molecules. 2019;24(11):2041. DOI: 10.3390/molecules24112041

[35] Beltrán G, Ruano MT, Jiménez A, Uceda M, Aguilera MP. Evaluation of virgin olive oil bitterness by total phenol content analysis. Eur J Lipid Sci Technol. 2007;109(3):193-197. DOI: 10.1002/ejlt.200600231

[36] Borges TH, Cabrera-Vique C, Seiquer I. Antioxidant properties of chemical extracts and bioaccessible fractions obtained from six Spanish monovarietal extra virgin olive oils: Assays in Caco-2 cells. Food Funct. 2015;6(7):2375-2383.

[37] Franco MN, Galeano-díaz T, Sánchez J, Miguel C De, Martín-vertedor D. Total phenolic compounds and tocopherols profiles of seven olive oil varieties grown in the south-west of Spain. J Oleo Sci. 2014;63(2):115-125.

[38] Borges TH, Serna A, López LC, Lara L, Nieto R, Seiquer I. Composition and antioxidant properties of Spanish extra virgin olive oil regarding cultivar, harvest year and crop stage. Antioxidants. 2019;8(7):217. DOI: 10.3390/antiox8070217

[39] Borges TH, Pereira JA, Cabrera-Vique C, Seiquer I. Study of the antioxidant potential of Arbequina extra virgin olive oils from Brazil and Spain applying combined models of simulated digestion and cell culture markers. J Funct Foods. 2017;37:209-218. DOI: 10.1016/j.jff.2017.07.059

[40] Sicari V, Leporini M, Giuffré AM, Aiello F, Falco T, Pagliuso MT, et al. Quality parameters, chemical compositions and antioxidant activities of Calabrian (Italy) monovarietal extra virgin olive oils from autochthonous (Ottobratica) and allochthonous (Coratina, Leccino, and Nocellara Del Belice) varieties. J Food Meas Charact. 2021;15(1):363-375. DOI: 10.1007/s11694-020-00640-9

[41] Bayram B, Esatbeyoglu T, Schulze N, Ozcelik B, Frank J, Rimbach G. Comprehensive analysis of polyphenols in 55 extra virgin olive oils by HPLC-ECD and their correlation

with antioxidant activities. Plant Foods Hum Nutr. 2012;67(4):326-336. DOI: 10.1007/s11130-012-0315-z

[42] Loizzo MR, Di Lecce G, Boselli E, Menichini F, Frega NG. Radical scavenging, total antioxidant capacity, and antiproliferative activity of phenolic extracts from extra virgin olive oil by Cultivar "Frantoio." Int J Food Prop. 2012;15(6):1345-1357. DOI: 10.1080/10942912.2010.522291

[43] Leporini M, Loizzo MR, Tenuta MC, Falco T, Sicari V, Pellicanò TM, et al. Calabrian extra-virgin olive oil from Frantoio cultivar: Chemical composition and health properties. Emirates J Food Agric. 2018;30(7):631-637. DOI: 10.9755/ejfa.2018.v30.i7.1743

[44] Kalogeropoulos N, Kaliora AC, Artemiou A, Giogios I. Composition, volatile profiles and functional properties of virgin olive oils produced by two-phase vs three-phase centrifugal decanters. LWT - Food Sci Technol. 2014;58(1):272-9. DOI: 10.1016/j.lwt.2014.02.052

[45] Bouaziz M, Jemai H, Khabou W, Sayadi S. Oil content, phenolic profiling and antioxidant potential of Tunisian olive drupes. J Sci Food Agric. 2010;90(10):1750-1758. DOI: 10.1002/jsfa.4013

[46] Nakbi A, Issaoui M, Dabbou S, Koubaa N, Echbili A, Hammami M, et al. Evaluation of antioxidant activities of phenolic compounds from two extra virgin olive oils. J Food Compos Anal. 2010;23:711-5. DOI: 10.1016/j.jfca.2010.05.003

[47] Gouvinhas I, Machado J, Gomes S, Lopes J, Martins-Lopes P, Barros AIRNA. Phenolic composition and antioxidant activity of monovarietal and commercial portuguese olive oils. J Am Oil Chem Soc. 2014;91(7):1197-1203.

[48] Kelebek H, Kesen S, Selli S. Comparative study of bioactive constituents in Turkish olive oils by LC-ESI/MS/MS. Int J Food Prop. 2015;18(10):2231-2245. DOI: 10.1080/10942912.2014.968788

[49] Köseoğlu O, Sevim D, Kadiroğlu P. Quality characteristics and antioxidant properties of Turkish monovarietal olive oils regarding stages of olive ripening. Food Chem. 2016;212:628-634. DOI: 10.1016/j.foodchem.2016.06.027

[50] Bajoub A, Carrasco-Pancorbo A, Ajal EA, Beltrán Maza G, Fernández-Gutiérrez A, Ouazzani N. Contribution to the establishment of a protected designation of origin for Meknès virgin olive oil: A 4-years study of its typicality. Food Res Int. 2014;66:332-343. DOI: 10.1016/j.foodres.2014.09.021

[51] El Qarnifa S, El Antari A, Hafidi A. Effect of maturity and environmental conditions on chemical composition of olive oils of introduced cultivars in Morocco. J Food Qual. 2019;2019. DOI: 10.1155/2019/1854539

[52] Olmo-García L, Polari JJ, Li X, Bajoub A, Fernández-Gutiérrez A, Wang SC, et al. Study of the minor fraction of virgin olive oil by a multi-class GC–MS approach: Comprehensive quantitative characterization and varietal discrimination potential. Food Res Int. 2019;125:108649. DOI: 10.1016/j.foodres.2019.108649

[53] Bajoub A, Pacchiarotta T, Hurtado-fernández E, Olmo-garcía L, García-Villalba R, Fernández-Gutiérrez A, et al. Comparing two metabolic profiling approaches (liquid chromatography and gas chromatography coupled to mass spectrometry) for extra-virgin olive oil phenolic compounds analysis : A botanical classification perspective. J Chromatogr A. 2016;1428:267-279. DOI: 10.1016/j.chroma.2015.10.059

[54] Mohamed M Ben, Rocchetti G, Montesano D, Ali S Ben, Guasmi F, Grati-Kamoun N, et al. Discrimination of Tunisian and Italian extra-virgin olive oils according to their phenolic and sterolic fingerprints. Food Res Int. 2018;106:920-927. DOI: 10.1016/j.foodres.2018.02.010

[55] Nescatelli R, Bonanni RC, Bucci R, Magrì AL, Magrì AD, Marini F. Geographical traceability of extra virgin olive oils from Sabina PDO by chromatographic fingerprinting of the phenolic fraction coupled to chemometrics. Chemom Intell Lab Syst. 2014;139:175-180.

[56] Romero M, Motilva M. Effect of climatic conditions on quality of virgin olive oil. In: Preedy VR, Watson RR, editors. Olives and Olive Oil in Health and Disease Prevention. Elsevier Inc.; 2010. p.43-50.

[57] Ben Rouina Y, Zouari M, Zouari N, Ben Rouina B, Bouaziz M. Olive tree (*Olea europaea* L. cv. Zelmati) grown in hot desert climate : Physio- biochemical responses and olive oil quality. Sci Hortic. 2020;261:108915. DOI: 10.1016/j.scienta.2019.108915

[58] Caruso G, Gucci R, Urbani S, Esposto S, Taticchi A, Di Maio I, et al. Effect of different irrigation volumes during fruit development on quality of virgin olive oil of cv. Frantoio. Agric Water Manag. 2014;134:94-103. DOI: 10.1016/j.agwat.2013.12.003

[59] Issaoui M, Flamini G, Brahmi F, Dabbou S, Ben Hassine K, Taamali A, et al. Effect of the growing area conditions on differentiation between Chemlali and Chétoui olive oils. Food Chem. 2010;119(1):220-225.

[60] Morales-Sillero A, Jiménez R, Fernández JE, Troncoso A, Beltrán G. Influence of fertigation in "Manzanilla de Sevilla" olive oil quality. HortScience. 2007;42(5):1157-1162. DOI: 10.21273/HORTSCI.42.5.1157

[61] Dag A, Erel R, Kerem Z, Ben-Gal A, Stern N, Bustan A, et al. Effect of nitrogen availability on olive oil quality. Acta Hortic. 2018;1199:465-469. DOI: 10.17660/ActaHortic.2018.1199.74

[62] Alagna F, Mariotti R, Panara F, Caporali S, Urbani S, Veneziani G, et al. Olive phenolic compounds: Metabolic and transcriptional profiling during fruit development. BMC Plant Biol. 2012;12:162. DOI: 10.1186/1471-2229-12-162

[63] Malheiro R, Rodrigues N, Pereira JA. Olive oil phenolic composition as affected by geographic origin, olive cultivar, and cultivation systems. In: Boskou D, editor. Olive and Olive Oil Bioactive Constituents. American Oil Chemists' Society Press; 2015. p.93-121. DOI: 10.1016/B978-1-63067-041-2.50010-0

[64] Romero N, Saavedra J, Tapia F, Sepúlveda B, Aparicio R. Influence of agroclimatic parameters on phenolic and volatile compounds of Chilean virgin olive oils and characterization based on geographical origin, cultivar and ripening stage. J Sci Food Agric. 2016;96(2):583-592. DOI: 10.1002/jsfa.7127

[65] Kalogeropoulos N, Kaliora AC. Effect of fruit maturity on olive oil phenolic composition and antioxidant capacity. In: Boskou D, editor. Olive and Olive Oil Bioactive Constituents. American Oil Chemists' Society Press; 2015. p.123-145.

[66] Servili M, Taticchi A, Esposto S, Urbani S, Selvaggini R, Montedoro G. Influence of the decrease in oxygen during malaxation of olive paste on the composition of volatiles and phenolic compounds in virgin olive oil. J Agric Food Chem. 2008;56(21):10048-10055. DOI: 10.1021/jf800694h

[67] Frankel E, Bakhouche A, Lozano-Sánchez J, Segura-Carretero A,

Fernández-Gutiérrez A. Literature review on production process to obtain extra virgin olive oil enriched in bioactive compounds. Potential use of byproducts as alternative sources of polyphenols. J Agric Food Chem. 2013;61(22):5179-5188.

[68] Antonini E, Farina A, Scarpa ES, Frati A, Ninfali P. Quantity and quality of secoiridoids and lignans in extra virgin olive oils : The effect of two- and three-way decanters on Leccino and Raggiola olive cultivars. Int J Food Sci Technol. 2015;67(1):9-15. DOI: 10.3109/09637486.2015.1121473

[69] Guerrini L, Migliorini M, Giusti M, Parenti A. The influence of crusher speed on extra virgin olive oil characteristics. Eur J Lipid Sci Technol. 2017;119(1):1600156. DOI: 10.1002/ejlt.201600156

[70] Ouni Y, Guerfel M, Chedly A, Salma Nayet M, Zarrouk M, Flamini G. Changes in volatile compounds and oil quality with malaxation time of Tunisian cultivars of *Olea europea*. Int J Food Sci Technol. 2013;48(1):74-81.

[71] Jolayemi OS, Tokatli F, Ozen B. Effects of malaxation temperature and harvest time on the chemical characteristics of olive oils. Food Chem. 2016;211:776-783. DOI: 10.1016/j.foodchem.2016.05.134

[72] Lolis A, Badeka A V., Kontominas MG. Quality retention of extra virgin olive oil , Koroneiki cv. packaged in bag-in- box containers under long term storage : A comparison to packaging in dark glass bottles. Food Packag Shelf Life. 2020;26:100549. DOI: 10.1016/j.fpsl.2020.100549

[73] Li X, Zhu H, Shoemaker CF, Wang SC. The effect of different cold storage conditions on the compositions of extra virgin olive oil. J Am Oil Chem Soc. 2014;91(9):1559-1570. DOI: 10.1007/s11746-014-2496-0

[74] Olmo-García L, Carrasco-Pancorbo A. Chromatography-MS based metabolomics applied to the study of virgin olive oil bioactive compounds: Characterization studies, agro-technological investigations and assessment of healthy properties. TrAC - Trends Anal Chem. 2021;135:116153. DOI: 10.1016/j.trac.2020.116153

[75] Wang P, Sun J, Zhang T, Liu W. Vibrational spectroscopic approaches for the quality evaluation and authentication of virgin olive oil. Applied Spectroscopy Reviews. 2016;51(10): 763-790. DOI: 10.1080/05704928.2016.1176034

[76] Lioupi A, Nenadis N, Theodoridis G. Virgin olive oil metabolomics: A review. J Chromatogr B. 2020;1150:122161. DOI: 10.1016/j.jchromb.2020.122161

[77] Waterhouse AL. Determination of Total Phenolics. Current Protocols in Food Analytical Chemistry. 2003;6:I1.1.1-I1.1.8. DOI: 10.1002/0471142913.fai0101s06

[78] Din I, Seccia S, Senatore A, Coppola D, Morelli E. Development and validation of an analytical method for total polyphenols quantification in extra virgin olive oils. Food Anal Methods. 2020;13:457-464. DOI: 10.1007/s12161-019-01657-7

[79] Dais P, Hatzakis E. Quality assessment and authentication of virgin olive oil by NMR spectroscopy: A critical review. Anal Chim Acta. 2013;765:1-27.

[80] Hatzakis E. Nuclear Magnetic Resonance (NMR) Spectroscopy in Food Science: A Comprehensive Review. Compr Rev Food Sci Food Saf. 2019;18(1):189-220. DOI: 10.1111/1541-4337.12408

[81] Ramautar R, Somsen GW, de Jong GJ. CE-MS for metabolomics:

Developments and applications in the period 2012-2014. Electrophoresis. 2015;36(1):212-224. DOI: 10.1002/elps.201400388

[82] Valdés A, Cifuentes A, León C. Foodomics evaluation of bioactive compounds in foods. TrAC - Trends Anal Chem. 2017;96:2-13. DOI: 10.1016/j.trac.2017.06.004

[83] Rigacci S. Olive oil phenols as promising multi-targeting agents against Alzheimer's disease. Adv Exp Med Biol. 2015;863:1-20. DOI: 10.1007/978-3-319-18365-7_1

[84] Rigacci S, Guidotti V, Bucciantini M, Nichino D, Relini A, Berti A, et al. Aβ(1-42) aggregates into non-toxic amyloid assemblies in the presence of the natural polyphenol oleuropein aglycon. Curr Alzheimer Res. 2011;8(8):841-852. DOI: 10.2174/156720511798192682

[85] Estruch R, Martínez-González MA, Corella D, Salas-Salvadó J, Ruiz-Gutiérrez V, et al. Effects of a Mediterranean-style diet on cardiovascular risk factors: A randomized trial. Ann Intern Med. 2006;145(1):1-11.

[86] Marcelino G, Hiane PA, Freitas K de C, Santana LF, Pott A, Donadon JR, et al. Effects of olive oil and its minor components on cardiovascular diseases, inflammation, and gut microbiota. Nutrients. 2019;11(8):1826. DOI: 10.3390/nu11081826

[87] Priore P, Caruso D, Siculella L, Gnoni G V. Rapid down-regulation of hepatic lipid metabolism by phenolic fraction from extra virgin olive oil. Eur J Nutr. 2015;54(5):823-833. DOI: 10.1007/s00394-014-0761-5

[88] Casaburi I, Puoci F, Chimento A, Sirianni R, Ruggiero C, Avena P, et al. Potential of olive oil phenols as chemopreventive and therapeutic agents against cancer : A review of *in vitro* studies. Mol Nutr Food Res. 2013;57:71-83.

[89] Lozano-Castellón J, López-Yerena A, De JFRA, Castillo-Alba JR del, Vallverdú-Queralt A, Elvira E-F, et al. Health-promoting properties of oleocanthal and oleacein: Two secoiridoids from extra-virgin olive oil. Crit Rev Food Sci Nutr. 2020;60(15):2532-2548. DOI: 10.1080/10408398.2019.1650715

[90] Tuck KL, Hayball PJ. Major phenolic compounds in olive oil: Metabolism and health effects. J Nutr Biochem. 2002;13:636-644.

[91] D'Archivio M, Filesi C, Varì R, Scazzocchio B, Masella R. Bioavailability of the polyphenols: Status and controversies. Int J Mol Sci. 2010;11(4):1321-1342.

[92] Fabiani R, Sepporta MV, Rosignoli P, De Bartolomeo A, Crescimanno M, Morozzi G. Anti-proliferative and pro-apoptotic activities of hydroxytyrosol on different tumour cells: The role of extracellular production of hydrogen peroxide. Eur J Nutr. 2012;51(4):455-464. DOI: 10.1007/s00394-011-0230-3

[93] St-Laurent-Thibault C, Arseneault M, Longpre F, Ramassamy C. Tyrosol and hydroxytyrosol, two main components of olive oil, protect N2a cells against amyloid-β-induced toxicity. Involvement of the NF-κB signaling. Curr Alzheimer Res. 2011;8(5):543-551. DOI: 10.2174/156720511796391845

[94] Cao K, Xu J, Zou X, Li Y, Chen C, Zheng A, et al. Hydroxytyrosol prevents diet-induced metabolic syndrome and attenuates mitochondrial abnormalities in obese mice. Free Radic Biol Med. 2014;67:396-407. DOI: 10.1016/j.freeradbiomed.2013.11.029

[95] Ozbek N, Bali EB, Karasu C. Quercetin and hydroxytyrosol attenuates xanthine/xanthine oxidase-induced toxicity in H9c2 cardiomyocytes by regulation of oxidative stress and stress-sensitive signaling pathways. Gen Physiol Biophys. 2015;34(4):407-414. DOI: 10.4149/gpb_2015021

[96] Muriana FJG, Montserrat-De La Paz S, Lucas R, Bermudez B, Jaramillo S, Morales JC, et al. Tyrosol and its metabolites as antioxidative and anti-inflammatory molecules in human endothelial cells. Food Funct. 2017;8(8):2905-2914. DOI: 10.1039/c7fo00641a

[97] Abe R, Beckett J, Abe R, Nixon A, Rochier A, Yamashita N, et al. Olive oil polyphenol oleuropein inhibits smooth muscle cell proliferation. Eur J Vasc Endovasc Surg. 2011;41(6):814-820. DOI: 10.1016/j.ejvs.2010.12.021

[98] Sepporta MV, Fuccelli R, Rosignoli P, Ricci G, Servili M, Morozzi G, et al. Oleuropein inhibits tumour growth and metastases dissemination in ovariectomised nude mice with MCF-7 human breast tumour xenografts. J Funct Foods. 2014;8(1):269-273. DOI: 10.1016/j.jff.2014.03.027

[99] Galanakis PA, Bazoti FN, Bergquist J, Markides K, Spyroulias GA, Tsarbopoulos A. Study of the interaction between the amyloid beta peptide (1-40) and antioxidant compounds by nuclear magnetic resonance spectroscopy. Biopolymers. 2011;96(3):316-327. DOI: 10.1002/bip.21558

[100] Mehraein F, Sarbishegi M, Aslani A. Evaluation of effect of oleuropein on skin wound healing in aged male Balb/c Mice. Cell J. 2014;16(1):25-30.

[101] Polini B, Digiacomo M, Carpi S, Bertini S, Gado F, Saccomanni G, et al. Oleocanthal and oleacein contribute to the in vitro therapeutic potential of extra virgin oil-derived extracts in non-melanoma skin cancer. Toxicol Vitr. 2018;52:243-250. DOI: 10.1016/j.tiv.2018.06.021

[102] Rosignoli P, Fuccelli R, Fabiani R, Servili M, Morozzi G. Effect of olive oil phenols on the production of inflammatory mediators in freshly isolated human monocytes. J Nutr Biochem. 2013;24(8):1513-1519. DOI: 10.1016/j.jnutbio.2012.12.011

[103] Czerwińska ME, Kiss AK, Naruszewicz M. Inhibition of human neutrophils NEP activity, CD11b/CD18 expression and elastase release by 3,4-dihydroxyphenylethanol-elenolic acid dialdehyde, oleacein. Food Chem. 2014;153:1-8. DOI: 10.1016/j.foodchem.2013.12.019

[104] Monti MC, Margarucci L, Riccio R, Casapullo A. Modulation of tau protein fibrillization by oleocanthal. J Nat Prod. 2012;75(9):1584-1588.

[105] Pei T, Meng Q, Han J, Li HSL, Song R, Sun B, et al. (−)-Oleocanthal inhibits growth and metastasis by blocking activation of STAT3 in human hepatocellular carcinoma. Oncotarget. 2016;7(28):43475-43491. DOI: 10.18632/oncotarget.9782

[106] Scotece M, Gómez R, Conde J, Lopez V, Gómez-Reino JJ, Lago F, et al. Further evidence for the anti-inflammatory activity of oleocanthal: Inhibition of MIP-1α and IL-6 in J774 macrophages and in ATDC5 chondrocytes. Life Sci. 2012;91(23-24):1229-1235. DOI: 10.1016/j.lfs.2012.09.012

[107] Cicerale S, Lucas L, Keast R. Biological activities of phenolic compounds present in virgin olive oil. Int J Mol Sci. 2010;11(2):458-479.

[108] Corona G, Tzounis X, DessÌ MA, Deiana M, Debnam ES, Visioli F, et al.

The fate of olive oil polyphenols in the gastrointestinal tract : Implications of gastric and colonic microflora-dependent biotransformation. Free Radic Res. 2006;40(6):647-658. DOI: 10.1080/10715760500373000

[109] Visioli F, Galli C, Grande S, Colonnelli K, Patelli C, Galli G, et al. Hydroxytyrosol excretion differs between rats and humans and depends on the vehicle of administration. J Nutr. 2003;133(8):2612-2615.

[110] Miró-Casas E, Covas M-I, Fitó M, Farré-Albadalejo M, Marrugat J, Torre R de la. Tyrosol and hydroxytyrosol are absorbed from moderate and sustained doses of virgin olive oil in humans. Eur J Clin Nutr. 2003;57:186-190.

[111] Manna C, Galletti P, Maisto G, Cucciolla V, D'Angelo S, Zappia V. Transport mechanism and metabolism of olive oil hydroxytyrosol in Caco-2 cells. FEBS Lett. 2000;470:341-344.

[112] Edgecombe SC, Stretch GL, Hayball PJ. Oleuropein, an antioxidant polyphenol from olive oil, is poorly absorbed from isolated perfused rat intestine. J Nutr. 2000;130(12):2996-3002.

[113] Boccio P Del, Deo A Di, Curtis A De, Celli N, Iacoviello L, Rotilio D. Liquid chromatography–tandem mass spectrometry analysis of oleuropein and its metabolite hydroxytyrosol in rat plasma and urine after oral administration. J Chromatogr B. 2003;785(1):47-56.

[114] Vissers MN, Zock PL, Roodenburg AJC, Leenen R, Katan MB. Olive oil phenols are absorbed in humans. J Nutr. 2002;132(3):409-417.

Chapter 9

Experimental Carcinogenesis with 7,12-Dimethylbenz(a)Anthrazene (DMBA) and Its Inhibition with Extra Virgin Olive Oil and a Diet of Mature Olives (*Picual* Variety)

Juan José Soto-Castillo and Isicio Ortega-Medina

Abstract

7,12-Dimethylbenz(a)anthrazene (DMBA) is a carcinogen that induces carcinomas within a few weeks of application. Forty-four male hamsters were divided into four groups: DMBA dissolved in paraffin oil (DMBA-PO), DMBA dissolved in olive oil (DMBA-OO), paraffin oil and olive oil. Their mouths were swabbed daily with paraffin oil or extra virgin olive oil alternatively for the first two weeks, during the biweekly application of DMBA at 0.5% diluted in paraffin oil or olive oil for five weeks and daily until the twentieth week. The animals in the DMBA-OO and olive oil groups received an additional diet of mature *Picual* olives. The DMBA-PO carcinogen effect (35 carcinomas) is 100% and the inhibitory effect 0. The use of olive oil as DMBA solvent and the *ad libitum* diet with *Picual* olive has an inhibitory effect of 80%, with only three intraepithelial carcinomas and four verrucous carcinomas occurring and no invasive carcinoma.

Keywords: DMBA, carcinogenesis, olive oil, chemoprevention, squamous cell carcinoma

1. Introduction

One of the first models of experimental carcinogenesis in animals was carried out by Salley in 1954 [1]. After applying various carcinogens, including 9,10-dimethyl-1,2 benzanthracene on the oral epithelium of Syrian hamsters for 3 months, Salley was able to verify the existence of squamous cell carcinomas (SCC) and lymphatic metastases. Subsequently, several authors have standardized this model and repeated it in order to achieve new knowledge about DMBA and the process of experimentally induced carcinogenesis [2, 3].

7,12-Dimethylbenz[a]anthracene (DMBA) is a polycyclic aromatic hydrocarbon which may, on its own, induce premalignant lesions and carcinomas within a few weeks after it is administered in mucosae [4, 5]. Commonly, it has been used in combination with ethanol as a promoter. DMBA is released after the combustion of tobacco -especially with cigarettes- or from animal fat when meat is grilled, and is also found in smoked meat and fish. This substance is, therefore, strongly involved in the carcinogenesis of oral, bronchiopulmonary and digestive tract malignancies [6–9].

In order to discover new drugs with cancer preventive effects, some authors have obtained promising outcomes at basic research level, specifically with substances such as salvinolic acid B [10] -derived from *Salvia miltiorrhiza*, used in fluorescence-, isothiocyanates [11] -synthetic derivatives of cabbages, squash, turnips and turnip greens-, *Buddleja incana* leaves, a tree that grows in Peru and Bolivia, Toona sinensis leaves [12], and olive oil extracts [9, 13–15]. In relation to the latter, and especially regarding its phenolic compounds, its antioxidant and cardiovascular protective properties are well known. In this sense, we have data stating that olive oil may act as preventive or inhibitor of carcinogenesis, and could even modify the nature of premalignant lesions that have already arisen, providing them a more benign and indolent behavior [14, 16].

2. Objective

To experimentally test the inhibitory effect on the carcinogenesis process of *Picual* variety extra virgin olive oil.

3. Methods

Forty-four male hamsters (*Syrian Golden*), 4-6 weeks old and weighing 60-80 g, were divided into four groups (two control and two experimental):

- Experimental DMBA-PO group (DMBA from Sigma Chemical Co.), 12 animals. The oral pouches were brushed daily with paraffin oil (PO) in the first two weeks. Then, a solution of 0.5% DMBA and PO was administered on Mondays and Fridays for five weeks; alternatively, PO was applied on Tuesdays, Wednesdays and Thursdays at the same time. Thereafter, animals received daily PO until the twentieth week. All hamsters were fed with standard feed, and *ad libitum* water (**Figure 1**).

- Experimental DMBA-OO group (DMBA from Sigma Chemical Co.), 12 animals. The oral pouches were brushed daily with OO in the first two weeks. Then, a solution of 0.5% DMBA and OO was administered on Mondays and Fridays every two days, for five weeks. Thereafter, animals received daily OO until the twentieth week. These hamsters were fed with standard feed, *Picual* variety ripe olives, and *ad libitum* water.

- Control DMBA-PO group, 10 hamsters. The oral pouches were brushed daily with PO for twenty weeks.

- Control DMBA-OO group, 10 hamsters. The oral pouches were brushed daily with extra virgin OO. Also, animals received diet with standard feed, *Picual* variety ripe olives, and *ad libitum* water for twenty weeks.

The animals in each group were sacrificed after twenty weeks. Then, a macroscopic description and histological analysis of the induced tumors in the oropharynx, esophagus and stomach were performed.

Figure 1.
The hamsters.

A carcinogenic effect of 100% was assigned to the total number of induced tumors in hamsters who received DMBA-PO. The inhibitory effect in the DMBA-OO group was established by the percentage difference over 100. An inhibitory effect >50% was considered significant in the DMBA-OO group.

This research work was examined and approved by the Ethical Committee for Animal Experimentation of the University of Seville (November 7, 2005). It met the requirements for experimentation with animals and was in accordance with the regulations in force in Spain and elsewhere the European Union.

4. Results

4.1 Macroscopically

The groups exposed to DMBA showed tumors of different characteristics. Nonspecific lesions and others more suggestive of malignancy were found in the oral pouches of the DMBA-OO group, with a predominance of the former. These findings included leukoplakia, denudation of the mucosa, ulcerations or tumors with a benign appearance. However, tumors in DMBA-OO group were less common and smaller than in DMBA-PO (**Figure 2A** and **B**).

On the other hand, the DMBA-PO group mostly showed malignant-looking neoplastic formations in the oral mucosa, such as ulcerated nodules, necrosis areas, exophytic and verrucous tumors, and areas with abundant vascularization.

In addition to the oral pouches, both DMBA groups presented tumors in the esophagus and stomach. Maximum and minimum measures of all lesions are shown in **Table 1**.

No visible lesions were found in the control groups which only received paraffin oil or olive oil.

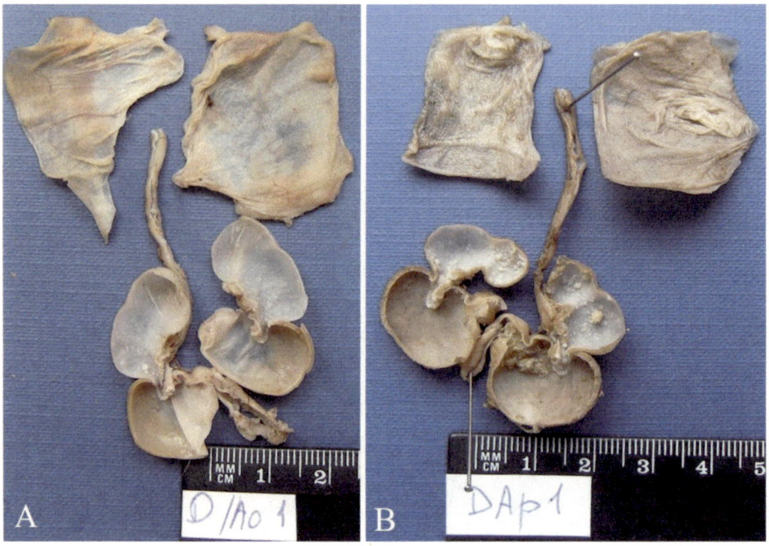

Figure 2.
Macroscopic comparison of the digestive tract of two animals belonging to the DMBA-OO group (A) and DMBA-PO group (B). (A) The oral pouches, esophagus and stomach showed few small lesions and benign appearance. (B) The oral pouches, esophagus, and stomach showed tissue retractions, larger tumors and apparently more malignant lesions.

Histological lesion	Location	Number of lesions		Mean diameter (Ø mm)	
		DMBA-PA	DMBA-OO	DMBA-PA	DMBA-OO
SQUAMOUS PAPILLOMA	Oral epithelium	1	11	1	0.14 (0.1-0.2)
	Esophagus	9	19	0.6 (0.2-1.2)	0.34 (0.2-1)
	Stomach	14	21	0.7 (0.3-1.2)	0.5 (0.2-1)
INTRAEPITHELIAL CARCINOMA	Oral epithelium	20	2	0.2 (0.1-0.4)	0.3 (0.3)
	Esophagus	0	1	0	0.3
	Stomach	1	0	0.6	0
VERRUCOUS CARCINOMA	Oral epithelium	1	2	1.7	4 (1-7)
	Esophagus	3	0	0.9 (0.8-1)	0
	Stomach	5	2	1.5 (0.8-2)	1.5 (1.2-1.8)
INVASIVE CARCINOMA	Oral epithelium	1	0	0.5	0
	Esophagus	2	0	3 (2-4)	0
	Stomach	2	0	5.3 (1.5-9)	0

Table 1.
Number, size and type of tumors in DMBA groups at 20th week.

4.2 Microscopically

The histological study at 20 weeks evidenced different types of lesions, demonstrating a complete carcinogenesis process in both DMBA groups: Squamous papillomas, intraepithelial carcinomas, verrucous carcinomas and invasive SCC.

SQUAMOUS PAPILLOMAS: Papillary projections lined with squamous epithelium were noted, showing hyperkeratosis and epithelial thickening. No atypia or mitotic activity was observed (**Figure 3**). Twelve papillomas were found among the groups exposed to DMBA (one in the DMBA-PO group and eleven in the DMBA-OO group). The differences regarding incidence of this kind of lesion were statistically significant (p .004).

INTRAEPITHELIAL CARCINOMAS (**Figure 4**): Twenty four intraepithelial carcinomas were identified. Twenty one occurred in the DMBA-PO group, and three in the DMBA-OO group. The differences observed between both groups were statistically significant (p .003).

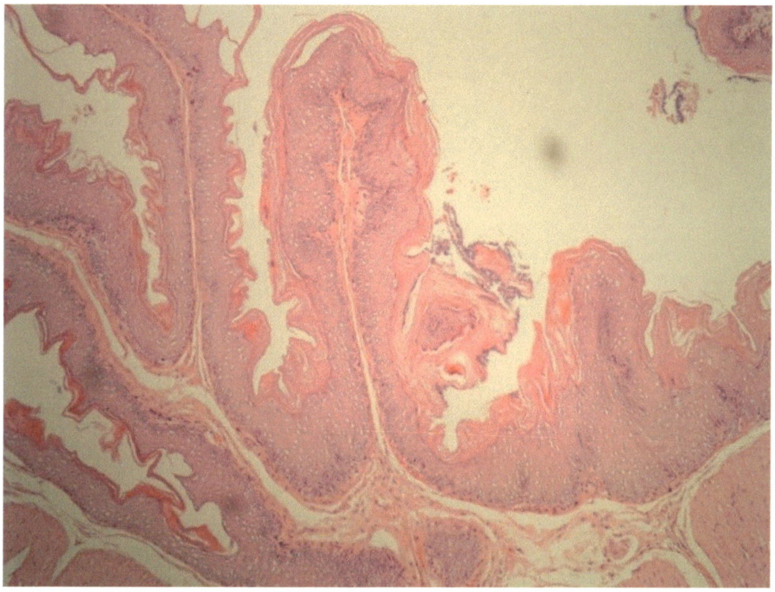

Figure 3.
Squamous papilloma is an exophytic lesion which shows typically papillary growth and highly differentiated epithelium.

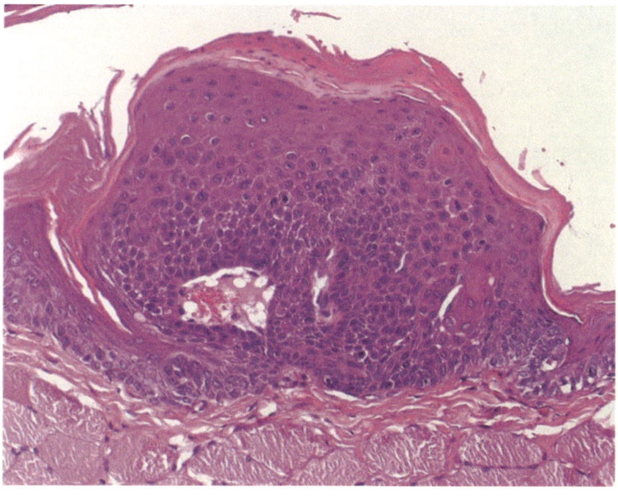

Figure 4.
Intraepithelial carcinoma is classically characterized by full-thickness with hyperkeratosis and parakeratosis, hypercellurality, nuclear atypia and mitotic figures. The epithelium-stroma interface is preserved.

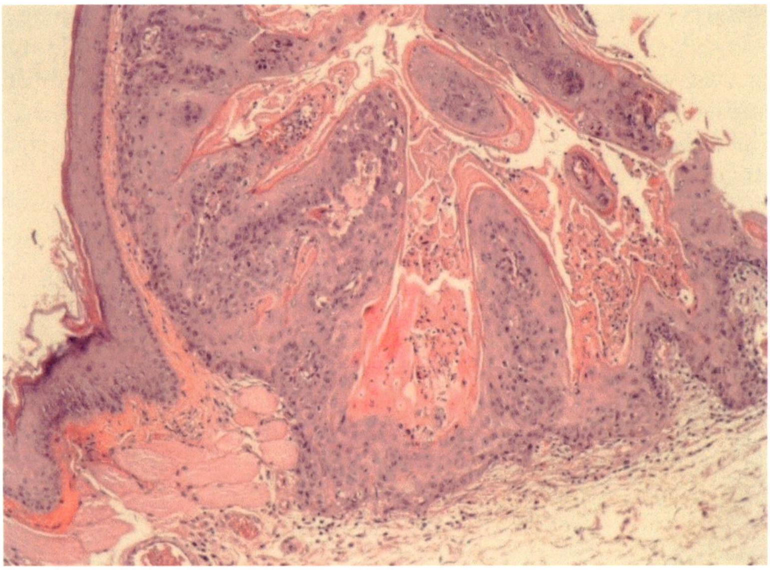

Figure 5.
SCC, verrucous carcinoma. Verrucous carcinoma is warty-appearing, highly differentiated, and shows hyperkeratosis. There is minimal atypia, abundant eosinophilic cytoplasm and normal mitotic figures. No invasion of the stroma by isolated neoplastic cells was observed.

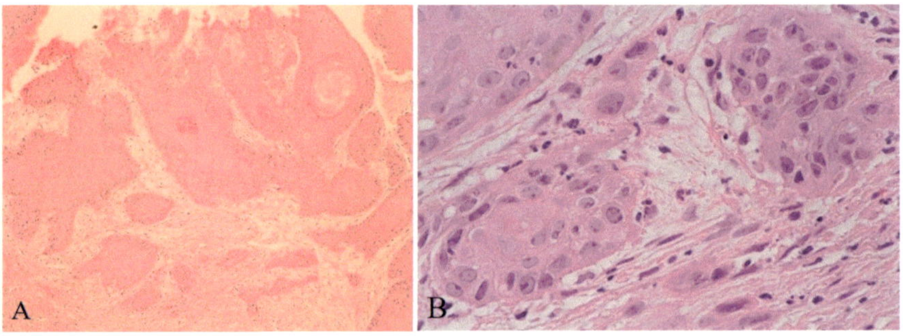

Figure 6.
Invasive SCC (A). The SCC is composed of infiltrating islands or nets of malignant cells, which form an irregular growth pattern (B).

Lesion		DMBA-PA (%)	DMBA-OO (%)
INTRAEPITHELIAL CARCINOMA	Carcinogenic effect	21 (100)	3 (14.3)
	Inhibitory effect		(85.7)
VERRUCOUS CARCINOMA	Carcinogenic effect	9 (100)	4 (44.4)
	Inhibitory effect		(55.6)
INVASIVE CARCINOMA	Carcinogenic effect	5 (100)	0
	Inhibitory effect		(100)
TOTAL CARCINOMAS	Carcinogenic effect	35 (100)	7 (20)
	Inhibitory effect		(80)

Table 2.
Carcinogenic and inhibitory effects of DMBA-PO/DMBA-OO, according to lesion subtypes.

VERRUCOUS CARCINOMAS: Several exophytic lesions with papillomatosis and infiltrative growth (**Figure 5**). Thirteen verrucous carcinomas were found, nine in the in the DMBA-PO group and four in the DMBA-OO group. This was not statistically significant (p .523).

INVASIVE CARCINOMAS (**Figure 6**): Light microscopy revealed epithelial proliferations that, like cords, invaded the adjacent stroma. In addition, the proliferating cells showed marked atypia and mitotic activity. Five invasive carcinomas were found in the DMBA-PO group.

The carcinogenic effect in the DMBA-PA group (35 carcinomas) corresponded to 100%, while in the DMBA-OO group (7 carcinomas), it was of 20%. According to the observed results, inhibitory effect seen in the DMBA-OO group was 86% for intraepithelial carcinoma, 56% for verrucous carcinoma, and 100% for SCC (**Table 2**).

No tumors were observed in the control animals.

5. Discussion

This research work about carcinogenesis is based on an experimental model of induced SCC after the administration of DMBA at 0.5% -dissolved in mineral oil- into the oral pouches of the hamster. We think, like Nagini and Kowshik [3], that the DMBA carcinogenesis model in hamster oral pouches is characteristic and highly representative of the "cancer induction". In addition, it is advantageous for the reproducibility of lesions, facilitates experimental research, and can be used as a test for chemotherapy and preventive agents. Also, in this work, the olive oil inhibitory effect on carcinogenesis has been studied alone -extra virgin olive oil applied before, during, and after DMBA, and *ad libitum* diet with ripe olives ripe of the *Picual* variety, from the olive harvest-, and combined -as a solvent for DMBA- [16].

The carcinomas produced in the upper gastrointestinal tract were SCC, similar to SCC of the oral mucosa in humans. These results coincide with those obtained in other experimental works [17, 18].

In oral carcinogenesis, using DMBA in hamsters, some authors have described the development of precancerous lesions and, subsequently, their progression towards intraepithelial carcinoma and invasive carcinoma after a few months. At 8 weeks, precancerous lesions usually appear. At 12 weeks, these evolve to intraepithelial carcinoma; eventually developing into invasive carcinomas at 18 weeks. This phenomenon, although slower, also occurs in humans [19]. The results obtained in our work resemble those of oral cancer progression described in the literature.

As in the field of experimental carcinogenesis, research on cancer prevention has continued to grow in recent decades, focusing on agents proposed for this purpose, although with few results yet. This is the case of the mediterranean diet, which is largely based on extra virgin olive oil, and that has been explored in the prevention of breast cancer [11, 13], and colorrectal cancer [9]. In the present work, the combination of olive oil as dissolvent, extra virgin olive oil applied before, during, and after DMBA application, and *ad libitum* diet with *Picual* variety olives, have been used as a preventive agent of DMBA carcinogenesis.

Menéndez et al. have shown that extra virgin olive oil polyphenols can inhibit erbB-2 malignant transformation of human breast cancer epithelial cells [14]. Owen et al. pointed out the importance of phenolic compounds isolated from olive oil as antioxidants and their anticancer potential [20].

In this sense, olive oil is composed of 99% different fatty acids, the most important being oleic acid, a monounsaturated fatty acid, with a richness of 60-80%, and other fatty acids -palmitic, stearic, palmitoleic, linoleic, and linolenic-. The remaining 1% is made up of vitamin E and natural antioxidants. The most important antioxidants are phenolic compounds, present in the mesocarp of the olive and in extra virgin olive oil, which are mainly responsible for the antioxidant properties and which are not present in any other vegetable oil. For this reason, the diet added to the standard feed that the hamsters received was ripe olives from the tree, recently harvested and not spoiled. The variety of olive richest in phenolic compounds is the *Picual* variety.

Keys et al. demonstrated an inversely proportional relationship between the incidence of cardiovascular diseases and the adoption of eating habits established in seven countries in the Mediterranean area [21]. It seems that this "cardiovascular protection" resides in the creation of an anti-atherosclerotic plasma profile, which is defined by a decrease in total cholesterol and low-density lipoprotein (LDL) cholesterol levels, as well as by an increase in high-density lipoprotein (HDL) cholesterol. Some studies have attributed these properties to the high content of oleic acid -monounsaturated grade acid of the omega-9 series- of olive oil [22].

Analyzing our results, we can affirm that the combination of olive oil as a solvent for DMBA, extra virgin olive oil applied before, during, and after DMBA administration, and *ad libitum* diet with *Picual* olives has shown the capability to reduce the malignant progression of lesions already started, and modify the malignant phenotype of some neoplasms, making it less aggressive.

It is possible that in the DMBA-OO group, -COOH groups and unsaturated bonds of the vegetable oil could absorb or react with carcinogen, decreasing the effective concentration of the carcinogen. The antioxidant effect and anticancer properties of extra virgin olive oil expressed by some authors are reinforced [18, 19].

The study of the lesions at 20 weeks showed a total of 59 neoplasms in the DMBA-PO group and 58 in the DMBA-OO group, so there were no differences in the absolute incidence of tumors. However, clear differences were observed regarding the type of neoplasms and malignancy. Eighty-eight percent of the tumors in the DMBA-OO group corresponded to benign squamous papilloma-type tumors, compared to 41% that developed in the DMBA-PO group; the rest were carcinomas.

In addition, hamsters that did not eat ripe olives and did not receive extra virgin olive oil, developed 21 intraepithelial carcinomas, 9 verrucous carcinomas, and 5 invasive carcinomas; while animals that received the olive oil as a solvent for DMBA, extra virgin olive oil -before, during, and after DMBA-, and *ad libitum* diet with *Picual* olive developed 3 intraepithelial carcinomas, 4 verrucous carcinomas, and no invasive squamous carcinoma.

6. Conclusions

The inhibitory effect of extra virgin olive oil (*Picual* variety) on the experimental chemical carcinogenesis is higher than 50% for carcinomas, especially for intraepithelial carcinoma and invasive squamous carcinoma.

Furthermore, the tumors originated in animals who received DMBA mixed with olive oil were predominantly benign, specifically of the squamous papilloma subtype.

Therefore, these data suggest that the extra virgin olive oil and the diet with ripe olives extracted from the harvesting of the tree may modulate the experimental carcinogenesis with DMBA, originating very well differentiated and not very aggressive tumors.

Conflict of interest

The authors declare no conflict of interest.

Author details

Juan José Soto-Castillo* and Isicio Ortega-Medina
Faculty of Medicine, University of Seville, Seville, Spain

*Address all correspondence to: jj27sc@gmail.com

References

[1] Salley JJ. Experimental carcinogenesis in the cheek pouch of the Syrian Hamster. J Dent Res. 1954;33: 253-62. DOI: 10.1177/00220345540330021201.

[2] Santis H, Shklar G, Chauncey HH. Histochemistry of experimentally induced leukoplakia and carcinoma of the hamster buccal pouch. Oral Surg Oral Med Oral Pathol. 1964;17:207-18. DOI: 10.1016/0030-4220(64)90144-6.

[3] Nagini S, Kowshik J. The hamster buccal pouch model of oral carcinogenesis. Methods Mol Biol. 2016;1422:341-50. DOI: 10.1007/978-1-4939-3603-8_29.

[4] Hassan MM, Shklar G, Solt D, Szabo G. Acute effect of DMBA application on mitotic activity of hamster buccal pouch epithelium. Oral Surg Oral Med Oral Pathol. 1985;59:491-8. DOI: 10.1016/0030-4220(85)90090-8.

[5] García FJG, Ortega VV, Sánchez NA, Jornet PL. Estudio comparativo de la aplicación del hidrocarburo aromático policíclico 7,12-dimetil-1,2-benzatraceno (DMBA) sobre la mucosa oral del hámster y del cobaya. Rev Española Patol. 2009;42: 287-95. DOI: 10.1016/S1699-8855(09)70196-5.

[6] Solt DB, Polverini PJ, Calderon L. Carcinogenic response of hamster buccal pouch epithelium to 4 polycyclic aromatic hydrocarbons. J Oral Pathol. 1987;16:294-302. DOI: 10.1111/j.1600-0714.1987.tb00697.x.

[7] Nagabhushan M, Ng YK, Elias R, Polverini PJ, Solt DB. Acute inhibition of DNA synthesis in hamster buccal pouch epithelium exposed to indirect acting carcinogens. Cancer Lett. 1990;53:163-73. DOI: 10.1016/0304-3835(90)90210-o.

[8] Calderon-Solt L, Solt DB. Gamma-glutamyl transpeptidase in precancerous lesions and carcinomas of oral, pharyngeal, and laryngeal mucosa. Cancer. 1985;56:138-43. DOI: 10.1002/1097-0142(19850701)56:1<138::aid-cncr2820560122>3.0.co;2-4.

[9] Bartolí R, Fernández-Bañares F, Navarro E, Castellà E, Mañé J, Alvarez M, et al. Effect of olive oil on early and late events of colon carcinogenesis in rats: Modulation of arachidonic acid metabolism and local prostaglandinE(2) synthesis. Gut. 2000;46:191-9. DOI: 10.1136/gut.46.2.191.

[10] Zhou ZT, Ge JP. The preventive effect of salvianolic acid B on malignant transformation of DMBA-induced oral premalignant lesion in hamsters. Carcinogenesis. 2006;27:826-32. DOI: 10.1093/carcin/bgi271.

[11] Warin R, Xiao D, Arlotti JA, Bommareddy A, Singh SV. Inhibition of human breast cancer xenograft growth by cruciferous vegetable constituent benzyl isothiocyanate. Mod Carcinog. 2010;49:500-7. DOI: 10.1002/mc.20600.

[12] Wang WC, Chen CY, Hsu HK, Lin LM, Chen YK. Chemopreventive effect of Toona sinensis leaf extracton 7,12-dimethylbenz(a)anthracene-induced hamster buccal pouch squamous cell carcinogenesis. Arch Oral Biol. 2016;70:130-42. DOI: 10.1016/j.archoralbio.2016.06.015.

[13] Manzanares MA, Solanas M, Moral R, Escrich R, Vela E, Costa I, et al. Dietary extra-virgin olive oil and corn oil differentially modulate the mRNA expression of xenobiotic-metabolizing enzimas in the liver and in the mammary gland in a rat chemical induced breast cancer model. Eur J Cancer Prev. 2015;24:215-22. DOI: 10.1097/CEJ.0000000000000032.

[14] Menéndez JA, Vázquez A, Oliveras C, García R, Carrasco A,

Fernández A, et al. Extra-virgin olive oil polyphenols inhibit HER2 (erbB-2)-induced malignant transformation in human breast epitelial cells: Relationship between the chemical structures of extra-virgin olive oil secoiridoids and lignans and their inhibitory activities on the tyrosine kinase activity of HER2. Int J Oncol. 2008;34:43-51.

[15] Solanas M, Hurtado A, Costa I, Moral R, Menéndez JA, Colomer R, et al. Effects of a high olive oil diet on the clinical behavior and histopathological features of rat DMBA-induced mammary tumors compared with a high corn oil diet. Int J Oncol. 2002;21:745-53.

[16] Soto-Castillo JJ, Ortega-Medina I. Carcinogénesis experimental con 7,12 dimetilbenzantraceno (DMBA) y su inhibición con aceite de oliva virgen extra y dieta con aceitunas maduras (variedad Picual). Rev Esp Patol. 2017:50(2):82-88. DOI: 10.1016/j.patol.2016.10.001.

[17] Chen YK, Lin LM. DMBA-induced hamster buccal pouch carcinoma and VX-induced rabbit cancer as a model for human oral carcinogenesis. Expert Rev Anticancer Ther. 2010;10:1485-96. DOI: 10.1586/era.10.108.

[18] Tang X-H, Knudsen B, Bemis D, Tickoo S, Gudas LJ. Oral cavity and esophageal carcinogenesis modeled in carcinogen-treated mice. Clin Cancer Res. 2004;10:301-13. DOI: 10.1158/1078-0432.ccr-0999-3.

[19] Vairaktaris E, Spyridonidou S, Papakosta V, Vylliotis A, Laza- ris A, Perrea D, et al. The hamster model of sequential oral oncogenesis. Oral Oncol. 2008;44:315-24. DOI: 10.1016/j.oraloncology.2007.08.015.

[20] Owen RW, Giacosa A, Hull WE, Haubner R, Spiegelhalder B, Bartsch H. The antioxidant/anticancer potential of phenolic compounds isolated from olive oil. Eur J Cancer. 2000;36:1235-47. DOI: 10.1016/s0959-8049(00)00103-9.

[21] Keys A, Menotti A, Karvonen MJ, Aravanis C, Blackburn H, Buzina R, et al. The diet and 15-year death rate in the seven countries study. Am J Epidemiol. 1986;124(6):903-15. DOI: 10.1093/oxfordjournals.aje.a114480.

[22] Ruiz V, Muriana FJ, Villar J. El aceite de oliva virgen y las enfermedades cardiovasculares. Perfil lipídico en plasma y composición lipídica de la membrana de eritrocito humano. Grasas y Aceites. 1998;49(1):9-29. DOI: 10.3989/gya.1998.v49.i1.703.